Rigging

By James Headley

A safety handbook for riggers, supervisors and other personnel who use rigging equipment to accomplish their work.

Rigging

by James Headley

A publication of:
Crane Institute of America Publishing & Products, Inc.,
3880 St. Johns Parkway
Sanford, FL 32771
e-mail: info@craneinstitute.com
web: www.craneinstitute.com

Fourteenth Edition. Copyright © 2013, 2012, 2011, 2010, 2009, 2007, 2006, 2005, 2004, 2002, 2001, 2000. Crane Institute of America Publishing & Products, Inc.

This publication is protected by U.S. and international copyright law. It may not be reproduced in whole or in part by any means (including electronic retrieval) without the express written permission of Crane Institute of America Publishing & Products, Inc. Violators will be pursued to the fullest extent under civil and criminal law.

Crane Institute of America Publishing & Products, Inc. has made strenuous efforts to ensure the reliability of the information provided in this publication. However, the publisher or author can accept no liability for inaccuracies or incompleteness. This publication should not be considered a substitute for relevant regulations and standards or manufacturers' specific recommendations, requirements or instructions.

The graphics used to illustrate information are not intended to be exact representations, but rather they are added to help the reader's understanding.

For information on other publications and products, or for details on Crane Institute's nationwide training and certification programs, call 1-800-832-2726. Outside USA call (407) 322-6800.

About the Author

James Headley has spent over 30 years working in the crane and rigging industry. After serving a crane apprenticeship through Operating Engineers, Local 312 in Birmingham, Alabama, he worked as a journeyman crane operator until entering the crane and rigging training and consulting business in 1984.

As president and director of Crane Institute of America, Inc., based near Orlando, Florida, Headley has developed and conducted training programs for dozens of major companies including ExxonMobil, Hanson Aggregates, Honeywell, Kennecott, Operating Engineers, Southern Company, Weyerhaeuser, U.S. Gypsum, USX, and military branches, etc. Headley has also been a featured speaker at crane and rigging conferences and is called upon frequently to investigate accidents and serve as an expert witness.

Headley is the author of numerous articles, the handbook "Mobile Cranes", and a set of field guides titled "Safe Craning". He also serves on the International Standards Organizations (ISO), Technical Committee SC5 and ANSI/ASME B30 main committee on cranes and lifting devices, and has served on the subcommittees B30.5 (Mobile and Locomotive Cranes), B30.9 (Slings) and B30.26 (Rigging Hardware). He received a two-year diploma in Machine Tool Technology from Bessemer State Technical Institute and a BS in Education from Southeastern Bible College in Birmingham, Alabama. He is married with three children and lives in Orlando, Florida.

Headley is also the CEO of Crane Institute of America Certification (CIC). CIC provides OSHA recognized and NCCA nationally accredited certifications for Crane Operators, Riggers, Signalpersons and Crane Inspectors. For more information, go to www.CICert.com.

Acknowledgements

No person achieves any measure of success without help and assistance from certain individuals and organizations along the way – certainly this has been true in my case. It is those who deserve special recognition that I would like to acknowledge.

Special thanks goes to John G. Watts for editing assistance, page design and layout, and many detailed illustrations which make the book much more clear and useful than it would have been otherwise, and for his patience in making the multitude of changes that were required. To Ann Campagnone and Stephanie Mandy for revision assistance.

I also want to thank the individuals who examined the manuscript for accuracy and offered valuable advice, much of which was used in this book. I am particularly grateful to:

- Mike Parnell of Industrial Training International for his help and advice on the advanced rigging examples.
- Ron Kohner of Landmark Engineering Services for helping clarify the math examples for dual crane lifts.

Contributions from the following entities are gratefully acknowledged:

- The Wire Rope Technical Board: for strength ratings of wire rope, capacity ratings for slings, additional technical information and illustrations.
- The Crosby Group: for technical information and reference art.
- Columbus Mckinnon Corp., for capacities and illustrations.

Finally, and most importantly, I would like to thank the people at Southeastern Bible College in Birmingham, Alabama, for providing me with an excellent education. Their emphasis on discipline, critical thinking and the pursuit of excellence has proven invaluable.

– James Headley

Rigging

This handbook very quickly should become an important asset for anyone working with rigging, no matter how much previous knowledge he or she has acquired on the subject already. The importance of using safe methods for lifting and moving heavy objects, and also for properly handling lighter items, is a subject that must be emphasized at every opportunity. There is no other sector of crane safety practices where a basic understanding of the fundamental principles is more crucial for safety and cost-efficiency.

A novice should study all of the graphical presentations in this handbook to form an initial perception of the various arrangements of common rigging and the different articles employed. An experienced rigger can also benefit from a review of the graphics, which portray unusually uniform and clear depictions of common practices. The advisories that accompany the graphics represent a compendium of advice that has evolved over a period of many years. Each of the instructions has been extracted from a variety of reputable sources. Each is aimed toward the best and most expeditious field practice, within the boundaries defined by maximum safety.

The emphasis on graphics is one of the best features of this handbook. As with its companion, "Mobile Cranes" (1999), the excellence of the computer-generated illustrations proves them to become vastly superior to old-fashioned photographs as teaching tools, particularly in a discussion between worker and supervisor.

At the turn of the century in the United States, all professionals engaged in rigging are encumbered by obsolete safety regulations promulgated by the federal government. These are three decades out of date. The path toward modernizing these regulations is blocked by our bureaucratic process, and our regulations do not cover some of the new techniques and materials readily available and in common use. An up-to-date handbook such as this one fills the information gap and educates today's rigger about current materials, methods, practices, and advisories. Everyone's safety is enhanced by it.

– Donald Sayenga

CONTENTS

WIRE ROPE 1-36

General Information 1
Classification 2
Construction 3-4
Grades 5
Cores 6
Lays 7-8
Special Ropes 9-10
    Rotation Resistant Ropes 9
    Compacted and Flattened Strand Ropes 10
Seizing and Cutting 11
Installation 12
Inspection 13-15
Measurement 16
Drum Spooling 17
Sheaves 18
Substitution 19
Lubrication 19
End Attachments 20-27
    Wire Rope Clips 20-24
    Wedge Sockets 25-27
Wire Rope Capacity Tables 28-36

SLINGS 37-94

General Information 37
Safe Use 38
Hitches 39-47
Sling Angles 48-49
Determining Sling Loading 50-57
Overlooked Sling Loading 58-59
Wire Rope Slings 60-68
    Sling Eyes 60
    D/d Ratio 61
    Hand Spliced Slings 62
    Inspection 63
    Wire Rope Sling Capacity Tables 64-68
Metal Mesh Slings 69-72
    Inspection 70
    Metal Mesh Sling Capacity Tables 71-72
Chain Slings 73-80

CONTENTS

SLINGS (Continued)
- Inspection 75-76
  - Chain Sling Capacity Tables 77-80
- Synthetic Web Slings 81-84
  - Inspection 83
  - Synthetic Web Sling Capacity Table 84
- Synthetic Roundslings 85-86
  - Inspection 85
  - Synthetic Roundsling Capacity Table 86
- Synthetic Rope Slings 87-94
  - Inspection 88
  - Synthetic Rope Sling Capacity Tables 89-94

HARDWARE **95-143**
- General Information 95
- Hooks 96-99
  - Application 97-98
  - Inspection 99
- Shackles 100-106
  - Application 101-104
  - Inspection 105
  - Capacity Table (Screw Pin Anchor, Bolt Anchor) 106
- Eye Bolts 107-115
  - Installation 108-111
  - Application 112-113
  - Inspection 114
  - Capacity Table (Forged, Shoulder Type) 115
- Hoist Rings 116-119
  - Installation 116
  - Application 117
  - Inspection 118
  - Capacity Table (UNC Threads) 119
- Master Links: Capacity Table (Alloy Steel) 120
- Turnbuckles 121-123
  - Inspection 122
  - Capacity Tables 123
- Blocks 124-129
  - Reeving and Mechanical Advantage 125
  - Single Part Load Line System 126
  - Angle Factors 127
  - Hoisting Loads 128-130

vii

CONTENTS

HARDWARE (Continued)
    Moving Loads Horizontally 131-134
        Inspection 135
Lever. 136
Lifting Beams 137-139
        Application. 138
        Inspection 139
Lever Operated Hoists 140-141
        Application. 140
        Inspection 141
Chain Hoists. 142-143
        Application. 142
        Inspection 143
Load Indicating Devices 144

PROCEDURES **145-181**

Power Lines 145-147
Determining Load Weight. 148-152
Center of Gravity 153-154
Handling Loads. 155-166
    Softeners. 155
    Attaching Unused Slings. 156
    Strength of Loads 157-158
    Improving Sling Efficiency. 158
    Turning Loads. 159
    Securing Loads. 160-161
    Tag Lines 162
    Fall Zone. 163
    Knots. 164-165
    Placement of Loads 166
Communicating with the Operator 167-173
    Voice Signals 168
    Special and Audible Signals 169
    Standard Hand Signals 170-173
Hoisting Personnel. 174-177
    Pre-Lift Considerations. 174
    Platform Specifications 175
    Selection of Rigging 176
    Trial and Test Lifts. 177
Multi-Crane Lifts 178
Dual Crane Lifts 179-181

CONTENTS

TRAINING AND CERTIFICATION/QUALIFICATION 182-184

SUPPORT SERVICES 185

OSHA ENDORSED ACCREDITED CERTIFICATIONS 186

    BASIC RIGGER/SIGNALPERSON. 186

    ADVANCED RIGGER. 186

    CRANE OPERATOR 186

    CRANE INSPECTOR/CERTIFIER 186

PRODUCTS 187-191

WIRE ROPE • General Information

The important and essential role that wire rope plays in hoisting and rigging cannot be underestimated. In fact, without wire rope it would be impossible to perform most if not all operations. Considering all the wire configurations, materials and close tolerances that make up wire rope, it is easy to understand why it is often referred to as a machine. It therefore is important that the user have at least a general understanding of wire rope in order to use it safely for its intended purpose.

Basic Components

Wire rope consists of three basic components: (1) *wires* that form the strand, (2) the multi-wire *strands,* and (3) a *core* of steel or fibers.

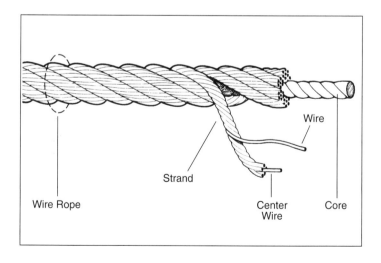

WIRE ROPE • Classification

Wire ropes are classified by a system consisting of two numbers (example: 6 x 19). The first number indicates the number of strands that make up the rope. The second number indicates the number of wires in a strand, which may or may not indicate the exact strand construction. For example, a 6 x 19 class rope could have as few as 15 wires and as many as 26.

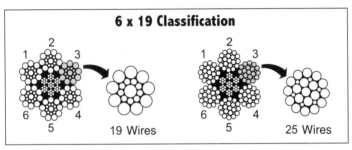

To avoid receiving the wrong wire rope it is best to order the specific rope construction required for the job.

Classification	Outer Strands	Wires per Strand	Max. No. Outer Wires
6 x 7	6	3–14	9
6 x 19	6	15–26	12
6 x 36	6	27–49	18
7 x 19	7	15–26	12
7 x 36	7	27–49	18
8 x 7	8	3–14	9
8 x 19	8	15–26	12
8 x 36	8	27–49	18
Rotation Resistant Ropes			
8 x 19	8	15–26	12
19 x 7	12	6–9	8
19 x 19	12	15–26	12
35 x 7	16–18	6–9	8
35 x 19	16–18	15–26	12

WIRE ROPE • Construction

Wire ropes are made with strands of various shapes, i.e., round strands, flattened strands. Most ropes used in hoisting and rigging operations are made up of round strands with the most common round strand constructions illustrated below.

Basic Round Strand Constructions

Single Layer: One layer of the same size wires laid around a center wire.

Filler Wire: Smaller wires are used to fill in the spaces between the larger wires.

Seale: Larger wires form outside of strand. Smaller wires are used on the inside of the strand.

Warrington: Outer wires are alternately large and small.

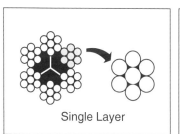
Single Layer

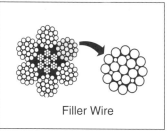
Filler Wire

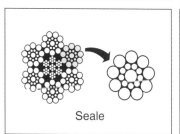
Seale

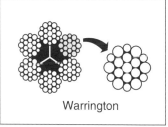
Warrington

WIRE ROPE • Construction

A combination of these basic strand constructions combined with the number and diameter of wires can be used to improve resistance to bending fatigue and abrasion.

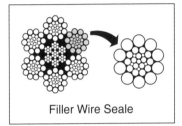

Filler Wire Seale

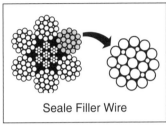

Seale Filler Wire

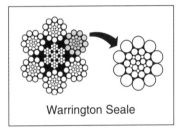

Warrington Seale

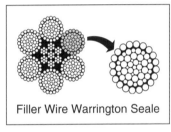

Filler Wire Warrington Seale

Generally, the greater the number of wires and the smaller in diameter they are, the more flexible the wire rope. Smaller outer wires have a greater ability to bend as they pass over sheaves and around drums, but less ability to resist abrasion. Conversely, ropes with larger outer wires have more resistance to abrasion, but do not bend as well.

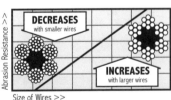

Size of Wires >>

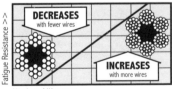

Number of Wires >>

WIRE ROPE • Grades

The strengths of wire ropes are basically determined by design and the material used to make the individual wires. There are a variety of materials used to make wire rope, but most hoisting and rigging ropes are made from high carbon steel. High carbon steel wire rope comes in a variety of grades. They are, in increasing tensile strength, improved plow steel (IPS), extra improved plow steel (EIPS) and extra extra improved plow steel (EEIPS). Wire rope made of EIPS is 15% stronger than IPS, while EEIPS ropes have 10% more strength than EIPS ropes.

Strength of Grades

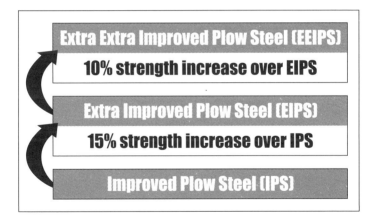

The most common finish for steel wire used to manufacture wire rope is uncoated or bright. Wire rope can also be galvanized with a zinc coating. Galvanized wires are normally 10% lower in strength than bright steel wires.

WIRE ROPE • Cores

The core provides the foundation about which the strands are laid. Cores are made up of either natural or man-made fibers, or steel. A steel core consists of either a strand or an independent wire rope. The three most commonly used core designations are: fiber core (FC), independent wire rope core (IWRC), and wire strand core (WSC).

Fiber core ropes are more bendable than steel core ropes, but crush more easily. Therefore, using fiber core ropes where they multi-wind on drums should be avoided. They also have about 8% less strength than steel cores and should not be used at temperatures greater than 180°.

Basic Cores

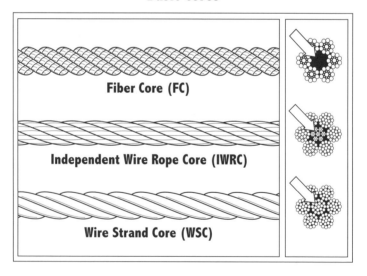

Fiber Core (FC)

Independent Wire Rope Core (IWRC)

Wire Strand Core (WSC)

WIRE ROPE • Lays

Strands that are laid around the core to the right are referred to as right lay rope. Conversely, strands that are laid in the opposite direction or to the left are called left lay rope. Right lay ropes are the most common wire ropes used in hoisting and rigging operations, whereas left lay ropes are only used in special applications.

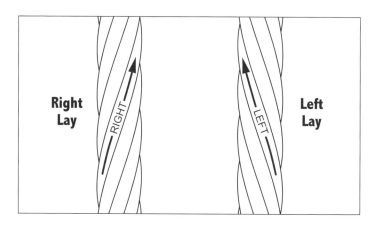

Lay Length

The term *lay length* refers to the distance it takes for a strand to make a complete revolution around the core.

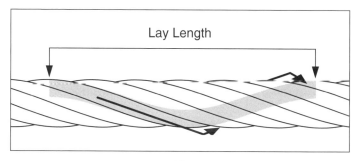

WIRE ROPE • Lays

Wire ropes are identified not only by their component parts and strand construction, but also by the direction of the strands and the outside wires of the strand. The direction of these strands and wires is referred to as *rope lay*. The lay of the wire rope affects its resistance to wear and flexibility; therefore use only rope recommended by the equipment or wire rope manufacturer.

Regular Lay

In regular lay ropes, wires appear to run parallel to the axis of the rope. As a commonly used multipurpose rope, it has good resistance to kinking, crushing and distortion.

Lang Lay

Wires in lang lay ropes are laid in the same direction as the strands, with the outside wires providing good resistance to abrasion. To prevent unwinding, lang lay ropes must have both ends attached and should never be used in a single part or used with a swivel.

Alternate Lay

Regular lay and lang lay strands are alternately laid around the core. These special ropes are usually used as boom hoist ropes in the crane industry.

WIRE ROPE • Special Ropes

Rotation Resistant Ropes

Rotation resistant ropes are a special category of wire rope designed to resist the tendency to spin or rotate under load. The most common rotation resistant rope has outside strands laid to the right and inside strands laid to the left. As one layer of strands attempts to rotate in one direction, it is counteracted by the other layer attempting to rotate in the opposite direction. These type ropes are easily damaged while in service and require special care and handling. Internal core slippage is one of the major problems that can occur.

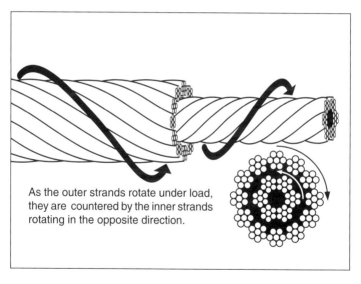

As the outer strands rotate under load, they are countered by the inner strands rotating in the opposite direction.

Some rotation resistant wire ropes are recommended to be reeved only with single lines. Open swivels should not be used with rotation resistant ropes of the 8 x 19, 19 x 7 and 19 x 19 classifications. Consult the rope manufacturer regarding the use of swivels with other multistrand constructions.

WIRE ROPE • Special Ropes

Compacted Strand Wire Rope

Compacted strand wire ropes are manufactured from strands which have been compacted or reduced in diameter prior to laying strands around the core. The compacting process flattens the outer wire surface and reforms the inner wires. This results in a smoother bearing surface of outer strands and an increase in strength over typical round strand ropes of the same diameter, because the metallic area is increased.

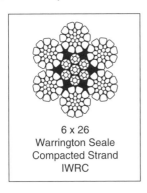

6 x 26
Warrington Seale
Compacted Strand
IWRC

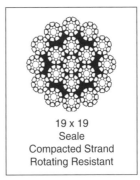

19 x 19
Seale
Compacted Strand
Rotating Resistant

Flattened Strand Wire Rope

Flattened strand wire ropes have exposed strand contours that are relatively flat. This flattened circular shape provides more contact area in sheave grooves, giving the outside wires greater resistance to wear compared to round strand ropes.

Flattened
Strand
IWRC

Flattened
Strand
FC

WIRE ROPE • Seizing and Cutting

Before cutting wire rope, proper seizing must be applied on both sides of the area to be cut. Failing to do this could result in the rope becoming distorted, flattened, or the strands loosened. For preformed rope one seizing on each side of the cut is sufficient. Non-preformed or rotation resistant ropes require a minimum of two seizings spread six diameters apart on each side of the cut.

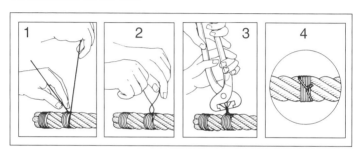

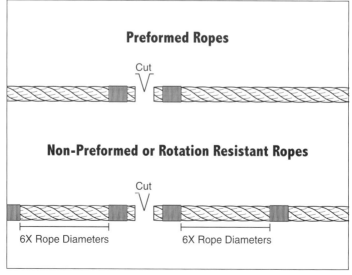

WIRE ROPE • Installation

When installing wire rope from reel to drum, it is important that the rope be wound from top to top or bottom to bottom and attached to the correct drum flange, both depending on how the drum winds (see illustrations below). Tension must also be maintained on the rope as it is wound on the drum, with each wrap wound tightly against the preceding wrap.

Improperly installing the wire rope can create numerous problems which can undermine the safety of lifting and moving a load. Therefore, extreme care must be applied.

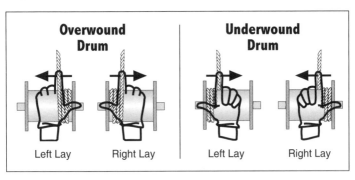

WIRE ROPE • Inspection

The inspection of wire rope is one of the most important requirements for maintaining the safe operating condition of hoisting and rigging equipment. Accordingly, all inspections must be performed by a qualified person.

For hoisting equipment the wire rope expected to be used must be inspected before beginning the operation. The entire rope, including attachments and end connections, must also be inspected monthly with a record kept and a comprehensive inspection with a record kept, annually.

If any of the following conditions exist (pages 13-15), the wire rope must be removed from service. Consult applicable standards for specific removal criteria.

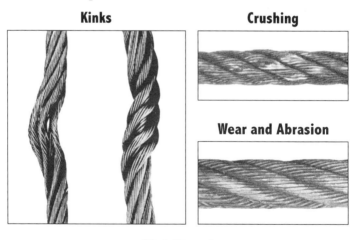

Kinks

Crushing

Wear and Abrasion

High Strand

WIRE ROPE • Inspection

Birdcaging

Core Protrusion

Wear of Outside Wires

Be sure to monitor the wire rope when outside wires are worn as this can result in broken wires and reduction in rope diameter beyond 5%. This would require that the rope be taken out of service.

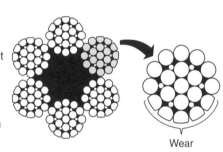

Wear

Reduction of Diameter

Reduction in rope diameter can be caused by: abrasion, corrosion, inner broken wires, rope stretch, loosening and tightening of rope lay, and core failure.

Broken Wires at Socket

Rope must be replaced when there is more than one broken wire at an end connection.

WIRE ROPE • Inspection

Broken Wires

Rope Replacement Based on Number of Broken Wires

Standard	Cranes and Equipment	Number Broken Wires In Running Ropes		Number Broken Wires In Standing Ropes	
		In One Rope Lay	In One Strand/Lay	In One Rope Lay	At End Connection
ANSI A10.4	Personnel Hoists	6*	3	2*	2
ASME B30.2	Overhead & Gantry Cranes	12*	4		
ASME B30.3	Construction Tower Cranes	12*	4	ns	3**
ASME B30.4	Portal, Tower, Pedestal Cranes	6*	3	ns	3**
ASME B30.5	Crawler, Locomotive & Truck Cranes	6*	3	3	2
ASME B30.6	Derricks	6*	3	3	2
ASME B30.7	Base Mounted Drum Hoists	6*	3	3	2
ASME B30.8	Floating Cranes & Derricks	6*	3	3	2

Standard	Equipment	Number Broken Wires In One Rope Lay	Number Broken Wires In One Strand/Lay
ANSI A10.5	Material Hoists	6*	ns
ASME B30.9	Slings (Wire Rope)	10	5

*also remove for one valley break   **also hoist ropes   ns = not specified

In running ropes. *ASME B30.16*: 6 randomly distributed broken wires in 6 rope diameters, or 3 broken wires in one strand in 6 rope diameters.

Rotation resistant rope replacement. *ASME B30.3, B30.5 and B30.16*: 2 randomly distributed broken wires in 6 rope diameters or 4 randomly distributed broken wires in 30 rope diameters. *ASME B30.4*: 4 randomly distributed broken wires in 1 lay, or 2 broken wires in 1 strand in 1 lay. *Other standards*: consult next update or rope manufacturer

In addition to the conditions shown, wire rope must be inspected for: heat damage; electric arc damage; improperly applied, severely corroded, cracked, bent or worn end connections; and contact with electrical sources.

WIRE ROPE • Measurement

All new wire rope should be measured or calibrated to ensure the correct diameter. New wire rope will normally be slightly larger than the nominal size. Once installed, all ropes must be measured periodically to make sure the diameter remains within allowable tolerances. The correct way to determine the diameter is to measure across the strand crowns and not across the flat areas of the strands. One way to do this is to turn the caliper around the rope.

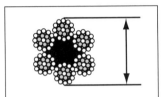

Must be taken out of service when there is more than a 5% reduction in nominal diameter.

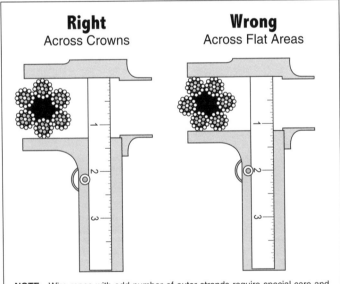

Right
Across Crowns

Wrong
Across Flat Areas

NOTE: Wire ropes with odd number of outer strands require special care and attention when measuring diameter.

WIRE ROPE • Drum Spooling

Once the wire rope has been properly installed, it must be observed frequently for proper spooling. If the rope is allowed to build up on one side of the drum it can fall off, causing the load to drop. It is required that at least two rope wraps remain on the drum at all times, but it is better for the entire first layer of rope to remain on the drum to act as a groove for the next layer.

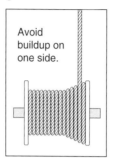

Avoid buildup on one side.

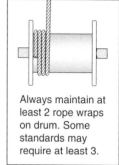

Always maintain at least 2 rope wraps on drum. Some standards may require at least 3.

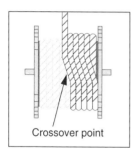

Crossover point

Pay particular attention to crossover points, as the wire rope is subject to severe abrasion and crushing while the rope is pushed over the two rope "grooves" and rides across the crown of the first rope layer. Crossover is unavoidable on the second and succeeding layers of wire rope.

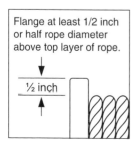

Flange at least 1/2 inch or half rope diameter above top layer of rope.

½ inch

To help prevent the wire rope from coming off the drum, standards require that the flange remain at least 1/2 inch or half the rope diameter above the top layer of rope, whichever is greater.

WIRE ROPE • Sheaves

In addition to inspecting the entire sheave for damage, use a sheave groove gauge to check the size, contour and amount of wear. If the sheave groove is too tight, the wire rope will be pinched or squeezed as it is forced into the groove. If the sheave groove is too loose, the rope will tend to flatten out. Also, inspect the groove surface for corrugation (imprint of the rope) and other abrasive defects. Damage such as described can accelerate degradation of the rope and shorten rope life.

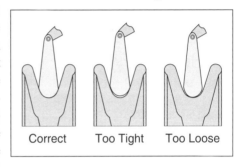

Correct   Too Tight   Too Loose

Sheaves should also turn easily and not wobble. Both conditions could be caused by broken or worn bushings or bearings.

The damage that is caused to wire rope when it comes out of the sheave groove is a major reason for rope breakage. To prevent this risk, sheave guards or keepers must be maintained (below right).

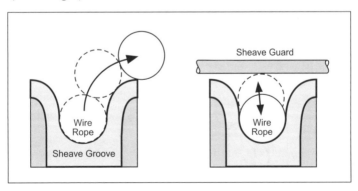

WIRE ROPE • Substitution

Replacing wire rope with a construction other than that specified by the rope or equipment manufacturer can create an unsafe condition. For example, it is a common practice to substitute a 6 x 19 classification rope with a 19 x 7 or 8 x 19 rotation resistant rope of the same diameter. Depending on the grade of steel, rope capacity can be reduced dramatically, even though the rotation resistant rope has the same diameter. Therefore, substitution must only be made with a rope that has a strength rating at least as great as the original rope furnished or recommended by the equipment manufacturer. Any change from the original size, grade or construction must be specified by the wire rope manufacturer or the equipment manufacturer.

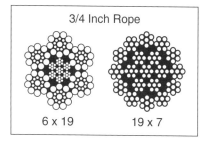

Lubrication

Although wire rope is lubricated during fabrication, it will require additional lubrication periodically after being put into service. Lubricants should be free of acids and alkalis, have sufficient adhesive strength, and be able to penetrate between the wires and strands. Used engine oil must always be avoided.

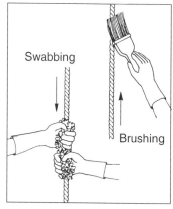

WIRE ROPE • End Attachments

Wire Rope Clips

There is a lot of wrong information in the field regarding the right way to install wire rope clips, the most common being that clips should be staggered during installation. Since more friction equals more efficiency, it stands to reason that the saddle, because of its larger surface area, must be installed on the live side of the rope. Having the saddle on the live side also greatly reduces the possibility of rope damage. A good way to know which side the saddle is to be installed is to remember, *never saddle a dead horse.*

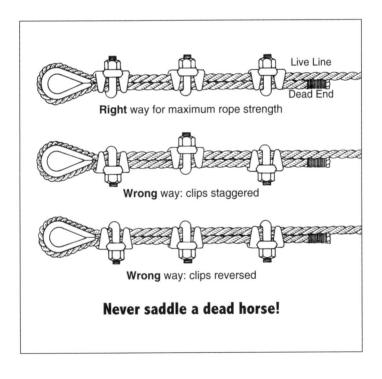

WIRE ROPE • End Attachments

Wire Rope Clips

Malleable clips are available, but only forged wire rope clips should be used for lifting loads. Malleable clips should only be used for light duty applications such as hand rails, fencing, guard rails, etc. To ensure that the end attachment develops the required strength, the following procedures should be followed during installation. The following is an example of the procedures to be followed when installing wire rope clips. Consult the clip manufacturer for specific procedures and before using clips above 400°F or below -40°F.

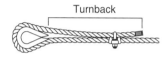

Step 1. Apply First Clip.
Turn back specified amount of rope from thimble or loop. Apply first clip one base width from dead end of rope. Apply U-Bolt over dead end of wire rope – live end rests in saddle. Also tighten nuts evenly, alternating until reaching the recommended torque.

Step 2. Apply Second Clip.
Apply the second clip as near the loop or thimble as possible. When <u>more than two clips are required</u>, apply the second clip as near the loop or thimble as possible, turn nuts on second clip firmly, but do not tighten. Proceed to Step 3.

Step 3. All Other Clips.
Space additional clips equally between the first two – take up rope slack – tighten nuts on each U-Bolt evenly, alternating from one nut to the other until reaching recommended torque.

Step 4. Test.
Apply first load of equal or greater weight expected to be used.

Step 5. Inspect and Recheck.
Inspect fitting and recheck torque on nuts frequently. Retighten to recommended torque if required.

WIRE ROPE • End Attachments

Wire Rope Clips

The preferred method of connecting two wire ropes together is to use interlocking turnback eyes with thimbles, using the recommended number of clips to form each eye.

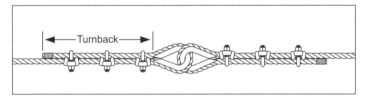

An alternate method is to use twice the number of clips as used for a turnback termination. The rope ends are placed parallel to each other, overlapping by twice the turnback amount shown in the application instructions. The minimum number of clips should be installed on each dead end. Spacing, installation torque, and other considerations still apply.

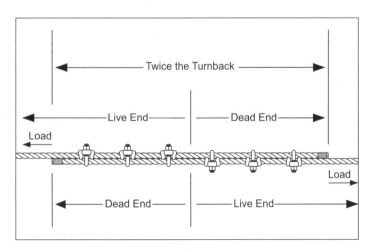

WIRE ROPE • End Attachments

U-Bolt Clips

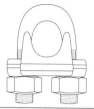

Clip Size, inches	Min # of Clips	Amount of Rope to Turnback, inches	Torque in lbs/ft	Weight, lbs/100
1/4	2	4 3/4	15	20
5/16	2	5 1/4	30	30
3/8	2	6 1/2	45	47
7/16	2	7	65	76
1/2	3	11 1/2	65	80
9/16	3	12	95	104
5/8	3	12	95	106
3/4	4	18	130	150
7/8	4	19	225	212
1	5	26	225	260
1 1/8	6	34	225	290
1 1/4	7	44	360	430
1 3/8	7	44	360	460
1 1/2	8	54	360	540
1 5/8	8	58	430	700
1 3/4	8	61	590	925
2	8	71	750	1300
2 1/4	8	73	750	1600

From the Crosby Group

- The number of clips is based upon using RRL or RLL wire rope, 6 x 19 or 6 x 37 class, FC or IWRC; IPS or XIP.
- Match the same size clip to the same size wire rope.
- If a pulley (sheave) is used for turning back the wire rope, add one additional clip.
- If a greater number of clips are used than shown, the amount of turnback should be increased proportionately.
- The tightening torque values shown are based upon the threads being clean, dry, and free of lubrication.
- Values do not apply to plastic coated wire rope.

WIRE ROPE • End Attachments

Double Saddle (Fist Grip) Clips

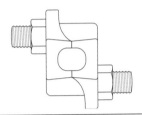

Clip Size, inches	Min # of Clips	Amount of Rope to Turnback, inches	Torque in lbs/ft	Weight, lbs/100
1/4	2	4	30	18
5/16	2	5	30	28
3/8	2	5 1/4	45	40
7/16	2	6 1/2	65	70
1/2	3	11 1/2	65	70
9/16	3	12 3/4	130	100
5/8	3	13 1/2	130	100
3/4	3	16	225	175
7/8	4	26	225	225
1	5	37	225	300
1 1/8	5	41	360	400
1 1/4	6	55	360	400
1 3/8	6	62	500	700
1 1/2	7	78	500	700

From the Crosby Group

- The minimum number of clips shown is based upon using RRL or RLL wire rope, 6 x 19 or 6 x 37 class, FC or IWRC; IPS or XIP.
- Match the same size clip to the same size wire rope.
- If a pulley (sheave) is used for turning back the wire rope, add one additional clip.
- If a greater number of clips are used than shown, the amount of turnback should be increased proportionately.
- The tightening torque values shown are based upon the threads being clean, dry, and free of lubrication.
- Values do not apply to plastic coated wire rope.

WIRE ROPE • End Attachments

Wedge Sockets

Wedge sockets are one of the most common end attachments found in the field because they are so easily installed and disassembled. However, to be used safely they must be installed properly. Make sure the socket, wedge and pin are the correct size for the wire rope used. Also, observe the condition of all components for damage and make sure the live line of the wire rope is aligned with the center line of the pin. Avoid using wedge sockets in temperatures above 400°F and below -4°F without first consulting the manufactuer.

WIRE ROPE • End Attachments

Wedge Sockets

Because of the small radius of wedge sockets and the possibility of internal core slippage, extreme care must be exercised when installing rotation resistant ropes. To prevent internal core slippage or loss of rope lay, ensure the dead end of rotation resistant rope is seized, welded or brazed before installation. Welding the dead end of standard wire rope is not recommended.

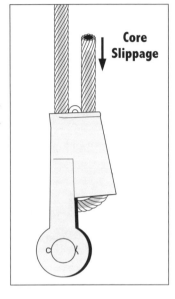

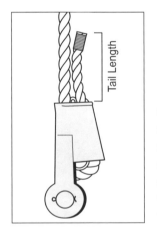

The *tail length* of the dead end of the wire rope (at left) should be a minimum of six rope diameters, but not less than six inches. For rotation resistant ropes the tail length of the dead end must be a minimum of twenty rope diameters, but not less than six inches.

WIRE ROPE • End Attachments

Wedge Sockets

There are a number of ways to correctly secure the dead end of a wire rope when installing a wedge socket. However, do not secure the dead end of the rope to the live rope, regardless of whether the saddle of the clip is placed on the dead end or live rope.

Attaching the dead end to the live line can damage, crimp or pinch the live line. This can also result in the load being transferred to the dead end. These conditions could ultimately result in the rope breaking unexpectedly at loads well below the rope's normal breaking strength.

Wrong

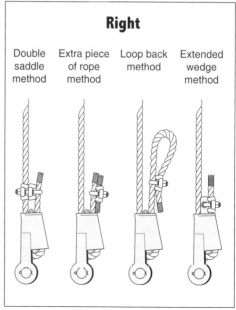

Right

Double saddle method | Extra piece of rope method | Loop back method | Extended wedge method

WIRE ROPE • 6 x 19 and 6 x 37 Classification

Bright (Uncoated) • IWRC

Nominal Diameter		Approximate Mass		Minimum Breaking Strength*					
				IPS**		EIPS**		EEIPS	
inches	mm	lb/ft	kg/m	tons	metric tons	tons	metric tons	tons	metric tons
3/8	9.5	0.26	0.39	6.56	5.95	7.55	6.85	8.3	7.54
7/16	11.5	0.35	0.52	8.89	8.07	10.2	9.25	11.2	10.18
1/2	13	0.46	0.68	11.5	10.4	13.3	12.1	14.6	13.3
9/16	14.5	0.59	0.88	14.5	13.2	16.8	15.2	18.5	16.7
5/8	16	0.72	1.07	17.9	16.2	20.6	18.7	22.7	20.6
3/4	19	1.04	1.55	25.6	23.2	29.4	26.7	32.4	29.4
7/8	22	1.42	2.11	34.6	31.4	39.8	36.1	43.8	39.7
1	26	1.85	2.75	44.9	40.7	51.7	46.9	56.9	51.6
1 1/8	29	2.34	3.48	56.5	51.3	65.0	59.0	71.5	64.9
1 1/4	32	2.89	4.30	69.4	63.0	79.9	72.5	87.9	79.8
1 3/8	35	3.50	5.21	83.5	75.7	96.0	87.1	106	95.8
1 1/2	38	4.16	6.19	98.9	89.7	114	103	125	113
1 5/8	42	4.88	7.26	115	104	132	120	146	132
1 3/4	45	5.67	8.44	133	121	153	139	169	153
1 7/8	48	6.5	9.67	152	138	174	158	192	174
2	52	7.39	11.0	172	156	198	180	217	198
2 1/8	54	8.35	12.4	192	174	221	200	243	220
2 1/4	57	9.36	13.9	215	195	247	224	272	246
2 3/8	60	10.4	15.5	239	217	274	249	301	274
2 1/2	64	11.6	17.3	262	238	302	274	332	301
2 5/8	67	12.8	19.0	288	261	331	300	364	330
2 3/4	70	14.0	20.8	314	285	361	327	397	360
2 7/8	74	15.3	22.8	341	309	392	356	431	392

* 1 ton = 2000 lbs. 1 metric ton = 2204 lbs.

* To convert to kilonewtons (kN), multiply tons (nominal strength) by 8.896; 1 lb = 4.448 newtons (N).

** Available with galvanized wires at strengths 10% lower than listed, or at equivalent strengths on special request.

WIRE ROPE • 6 x 19 and 6 x 37 Classification

Bright (Uncoated) • FC

Nominal Diameter		Approximate Mass		Minimum Breaking Strength*	
				IPS**	
inches	mm	lb/ft	kg/m	tons	metric tons
3/8	9.5	0.24	0.35	6.10	5.53
7/16	11.5	0.32	0.48	8.27	7.50
1/2	13	0.42	0.63	10.7	9.71
9/16	14.5	0.53	0.79	13.5	12.2
5/8	16	0.66	0.98	16.7	15.1
3/4	19	0.95	1.41	23.8	21.6
7/8	22	1.29	1.92	32.2	29.2
1	26	1.68	2.5	41.8	37.9
1 1/8	29	2.13	3.17	52.6	47.7
1 1/4	32	2.63	3.91	64.6	58.6
1 3/8	35	3.18	4.73	77.7	70.5
1 1/2	38	3.78	5.63	92.0	83.5
1 5/8	42	4.44	6.61	107	97.1
1 3/4	45	5.15	7.66	124	112
1 7/8	48	5.91	8.8	141	128
2	52	6.72	10.0	160	145
2 1/8	54	7.59	11.3	179	162
2 1/4	57	8.51	12.7	200	181
2 3/8	60	9.48	14.1	222	201
2 1/2	64	10.5	15.6	244	221
2 5/8	67	11.6	17.3	268	243
2 3/4	70	12.7	18.9	292	265

* 1 ton – 2000 lbs. 1 metric ton = 2204 lbs.

* To convert to kilonewtons (kN), multiply tons (nominal strength) by 8.896; 1 lb = 4.448 newtons (N).

** Available with galvanized wires at strengths 10% lower than listed, or at equivalent strengths on special request.

WIRE ROPE • 8 x 19 Classification

Rotation Resistant • Bright (Uncoated) • IWRC

Nominal Diameter		Approximate Mass		Minimum Breaking Strength*			
				IPS**		EIPS**	
inches	mm	lb/ft	kg/m	tons	metric tons	tons	metric tons
1/2	13	0.47	0.70	10.1	9.16	11.6	10.5
9/16	14.5	0.60	0.89	12.8	11.6	14.7	13.3
5/8	16	0.73	1.09	15.7	14.2	18.1	16.4
3/4	19	1.06	1.58	22.5	20.4	25.9	23.5
7/8	22	1.44	2.14	30.5	27.7	35.0	31.8
1	26	1.88	2.80	39.6	35.9	45.5	41.3
1 1/8	29	2.39	3.56	49.8	45.2	57.3	51.7
1 1/4	32	2.94	4.37	61.3	55.6	70.5	64.0
1 3/8	35	3.56	5.30	73.8	67.0	84.9	77.0
1 1/2	38	4.24	6.31	87.3	79.2	100.0	90.7

* 1 ton = 2000 lbs. 1 metric ton = 2204 lbs.

* To convert to kilonewtons (kN), multiply tons (minimum breaking strength) by 8.896; 1 lb = 4.448 newtons (N).

** Available with galvanized wires at strengths 10% lower than listed, or at equivalent strengths on special request.

NOTE: The given strengths for 8 x 19 rotation resistant ropes are applicable only when a test is conducted on a new rope fixed at both ends. When the rope is in use, and one end is free to rotate, the nominal strength is reduced.

WIRE ROPE • 19 x 7 Classification

Rotation Resistant • Bright (Uncoated)

Nominal Diameter		Approximate Mass		Minimum Breaking Strength*			
				IPS**		EIPS**	
inches	mm	lb/ft	kg/m	tons	metric tons	tons	metric tons
1/2	13	0.45	0.67	9.85	8.94	10.8	9.8
9/16	14.5	0.58	0.86	12.4	11.2	13.6	12.3
5/8	16	0.71	1.06	15.3	13.9	16.8	15.2
3/4	19	1.02	1.52	21.8	19.8	24.0	21.8
7/8	22	1.39	2.07	29.5	26.8	32.5	29.5
1	26	1.82	2.71	38.3	34.7	42.2	38.3
1 1/8	29	2.30	3.42	48.2	43.7	53.1	48.2
1 1/4	32	2.84	4.23	59.2	53.7	65.1	59.1
1 3/8	35	3.43	5.10	71.3	64.7	78.4	71.1
1 1/2	38	4.08	6.07	84.4	76.6	92.8	84.2

* 1 ton = 2000 lbs. 1 metric ton = 2204 lbs.

* To convert to kilonewtons (kN), multiply tons (minimum breaking strength) by 8.896; 1 lb = 4.448 newtons (N).

** Available with galvanized wires at strengths 10% lower than listed, or at equivalent strengths on special request.

NOTE: The given strengths for 8 x 19 rotation resistant ropes are applicable only when a test is conducted on a new rope fixed at both ends. When the rope is in use, and one end is free to rotate, the nominal strength is reduced.

WIRE ROPE • 6 x 19 and 6 x 37 Classification

Compacted Strand • Bright (Uncoated) • FC & IWRC

Nominal Dia.		Approximate Mass				Minimum Breaking Strength*			
inches	mm	lb/ft		kg/m		tons		metric tons	
		FC	IWRC	FC	IWRC	FC	IWRC	FC	IWRC
3/8	9.5	0.26	0.31	0.39	0.46	7.39	8.3	6.7	7.53
7/16	11.5	0.35	0.39	0.52	0.58	10.0	11.2	9.07	10.2
1/2	13	0.46	0.49	0.68	0.73	13.0	14.6	11.8	13.2
9/16	14.5	0.57	0.63	0.85	0.94	16.4	18.5	14.9	16.8
5/8	16	0.71	0.78	1.06	1.16	20.2	22.7	18.3	20.6
3/4	19	1.03	1.13	1.53	1.68	28.8	32.4	26.1	29.4
7/8	22	1.40	1.54	2.08	2.29	39.0	43.8	35.4	39.7
1	26	1.82	2.00	2.71	2.98	50.7	56.9	46.0	51.6
1 1/8	29	2.31	2.54	3.44	3.78	63.6	71.5	57.7	64.9
1 1/4	32	2.85	3.14	4.24	4.67	78.2	87.9	70.9	79.7
1 3/8	35	3.45	3.80	5.13	5.65	94.1	106	85.4	96.1
1 1/2	38	4.10	4.50	6.10	6.70	111	125	101	113
1 5/8	42	4.80	5.27	7.14	7.84	130	146	118	132
1 3/4	45	5.56	6.12	8.27	9.11	150	169	136	153
1 7/8	48	6.38	7.02	9.49	10.4	171	192	155	174
2	51	7.26	7.98	10.8	11.9	193	217	175	197

* 1 ton = 2000 lbs. 1 metric ton = 2204 lbs.

* To convert to kilonewtons (kN), multiply tons (nominal strength) by 8.896;
  1 lb = 4.448 newtons (N).

WIRE ROPE • 19 Strand Classification

Compacted Strand • Rotation Resistant Bright (Uncoated)

Nominal Diameter		Approximate Mass		Minimum Breaking Strength*			
				tons		metric tons	
inches	mm	lb/ft	kg/m	Standard	High Strength	Standard	High Strength
3/8	9.5	0.31	0.46	7.55	8.3	6.85	7.53
7/16	11.5	0.40	0.59	10.2	11.2	9.25	10.2
1/2	13	0.54	0.80	13.3	14.6	12.1	13.2
9/16	14.5	0.69	1.03	16.8	18.5	15.2	16.8
5/8	16	0.85	1.26	20.6	22.7	18.7	20.6
3/4	19	1.25	1.86	29.4	32.4	26.7	29.4
7/8	22	1.68	2.50	39.8	43.8	36.1	39.7
1	26	2.17	3.23	51.7	56.9	46.9	51.6
1 1/8	29	2.75	4.09	65.0	71.5	59.0	64.9
1 1/4	32	3.45	5.13	79.9	87.9	72.5	79.7
1 3/8	35	4.33	6.44	96.0	106.0	87.1	96.1
1 1/2	38	5.11	7.60	114.0	125.0	103.0	113.0

* 1 ton = 2000 lbs. 1 metric ton = 2204 lbs.

* To convert to kilonewtons (kN), multiply tons (nominal strength) by 8.896; 1 lb = 4.448 newtons (N).

WIRE ROPE • 6 x 19 and 6 x 37 Classification

Compacted (Swaged) • Bright (Uncoated) • IWRC

Nominal Diameter		Approximate Mass		Minimum Breaking Strength*	
				EIPS	
inches	mm	lb/ft	kg/m	tons	metric tons
1/2	13.0	0.55	0.82	15.5	14.0
9/16	14.5	0.70	1.04	19.6	17.8
5/8	16	0.87	1.29	24.2	22.0
3/4	19	1.25	1.86	34.9	31.7
7/8	22	1.70	2.53	47.4	43.0
1	26	2.22	3.30	62.0	56.3
1 1/8	29	2.80	4.16	73.5	66.7
1 1/4	32	3.40	5.05	90.0	81.8
1 3/8	35	4.20	6.24	106.0	96.2
1 1/2	38	5.00	7.43	130.0	118.0

* 1 ton = 2000 lbs. 1 metric ton = 2204 lbs.

* To convert to kilonewtons (kN), multiply tons (nominal strength) by 8.896; 1 lb = 4.448 newtons (N).

WIRE ROPE • 6 x 25 B, 6 x 27 H and 6 x 30 G

Flattened Strand • Bright (Uncoated) • IWRC

Nominal Diameter		Approximate Mass		Minimum Breaking Strength*			
				IPS**		EIPS**	
inches	mm	lb/ft	kg/m	tons	metric tons	tons	metric tons
1/2	13	0.47	0.70	12.6	11.4	14	12.7
9/16	14.5	0.60	0.89	16.0	14.5	17.6	16
5/8	16	0.73	1.09	19.6	17.8	21.7	19.7
3/4	19	1.06	1.58	28.1	25.5	31	28.1
7/8	22	1.46	2.17	38.0	34.5	41.9	38
1	26	1.89	2.83	49.4	44.8	54.4	49.4
1 1/8	29	2.39	3.56	62.2	56.4	68.5	62.1
1 1/4	32	2.95	4.39	76.3	69.2	84	76.2
1 3/8	35	3.57	5.31	91.9	83.4	101	91.6
1 1/2	38	4.25	6.32	108	98	119	108
1 5/8	42	4.98	7.41	127	115	140	127
1 3/4	45	5.78	8.60	146	132	161	146
1 7/8	48	6.65	9.90	167	152	184	167
2	52	7.56	11.3	189	171	207	188

* 1 ton = 2000 lbs. 1 metric ton = 2204 lbs.

* To convert to kilonewtons (kN), multiply tons (minimum breaking strength) by 8.896; 1 lb = 4.448 newtons (N).

** Available with galvanized wires at strengths 10% lower than listed, or at equivalent strengths on special request.

WIRE ROPE • 6 x 25 B, 6 x 27 H and 6 x 30 G

Flattened Strand • Bright (Uncoated) • FC

Nominal Diameter		Approximate Mass		Minimum Breaking Strength*	
				Improved Plow Steel**	
inches	mm	lb/ft	kg/m	tons	metric tons
1/2	13	0.45	0.67	11.8	10.8
9/16	14.5	0.57	0.85	14.9	13.5
5/8	16	0.70	1.04	18.3	16.6
3/4	19	1.01	1.50	26.2	23.8
7/8	22	1.39	2.07	35.4	32.1
1	26	1.80	2.68	46.0	41.7
1 1/8	29	2.28	3.39	57.9	52.5
1 1/4	32	2.81	4.18	71.0	64.4
1 3/8	35	3.40	5.06	85.5	77.6
1 1/2	38	4.05	6.03	101	91.6
1 5/8	42	4.75	7.07	118	107
1 3/4	45	5.51	8.20	138	123
1 7/8	48	6.33	9.42	155	141
2	52	7.20	10.70	176	160

* 1 ton = 2000 lbs. 1 metric ton = 2204 lbs.

* To convert to kilonewtons (kN), multiply tons (nominal strength) by 8.896; 1 lb = 4.448 newtons (N).

** Available with galvanized wires at strengths 10% lower than listed, or at equivalent strengths on special request.

SLINGS • General Information

There are a wide variety of slings available to riggers – all made for specific applications. They can be manufactured from fiber, wire rope, metal mesh, chain, or synthetic materials such as nylon, polypropylene and polyester. Slings can also be equipped with various components such as hooks, links and rings. They should be stored where they are protected from being damaged or degraded chemically.

Sling Types and Materials

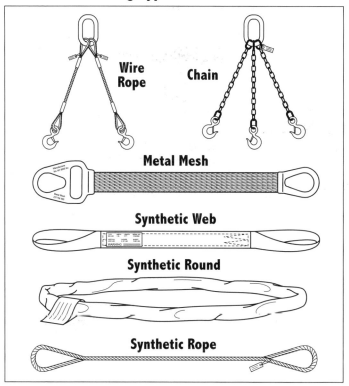

SLINGS • Safe Use

The safe use of slings, which requires staying within their rated capacity, largely depends on three important factors:
1. The *hitch* in which the sling is configured.
2. The *angle* of the sling.
3. The *sharpness* of the edges of the load which the sling passes around.

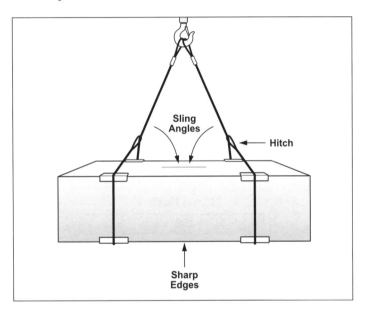

Note: the information contained on pages 38-59 applies to all types of slings: wire rope, metal mesh, alloy steel chain, synthetic web, synthetic roundslings, and synthetic rope. Information unique to a particular type of sling is addressed under its own heading in a subsequent section (see table of contents for pages 60-94).

SLINGS • Hitches

The method in which a sling is rigged or attached to a load is referred to as a *"hitch"*. The weight and shape of the load will largely determine which type of slings and hitches are used.

There are three basic types of hitches: vertical, choker, and basket, with each hitch capable of being set into various configurations.

Basic Hitches

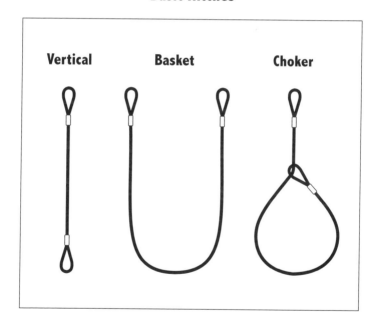

SLINGS • Hitches

Vertical Hitch

A sling is considered a vertical hitch when one end is attached to the load and the other end is attached to the lifting device or mechanism with the angle of loading being less than 5°. A vertical hitch should not be used for lifting loose material or loads that are difficult to balance. This type of hitch is best used with a shackle attached to an eye bolt or lifting eye.

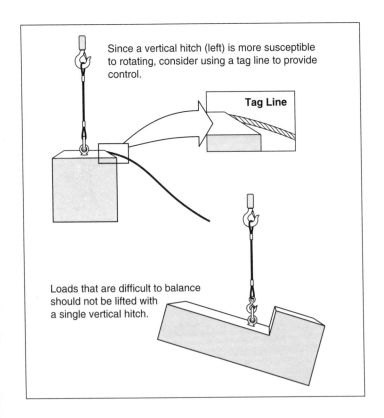

Since a vertical hitch (left) is more susceptible to rotating, consider using a tag line to provide control.

Tag Line

Loads that are difficult to balance should not be lifted with a single vertical hitch.

SLINGS • Hitches

Bridle Hitches

A bridle sling or hitch is composed of two or more individual legs attached to a lifting hook or gathered in a fitting. This hitch provides good load stability when the load weight is distributed among the legs and the hoisting hook is directly over the load's center of gravity.

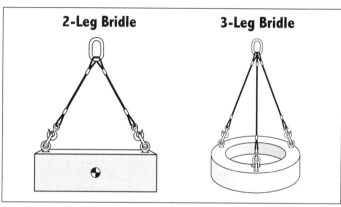

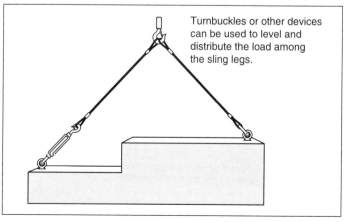

Turnbuckles or other devices can be used to level and distribute the load among the sling legs.

SLINGS • Hitches

Bridle Hitches

Unless all of the sling legs are the same length and equally spaced around the load's center of gravity, the loading on the legs of a 3- or 4-leg bridle sling may not be equal. In some situations when lifting and moving loads with a 3- or 4-leg bridle sling, 2 legs could end up carrying the load while the other leg(s) acts to balance it. In such cases, the capacity of two sling legs must be great enough to support the load.

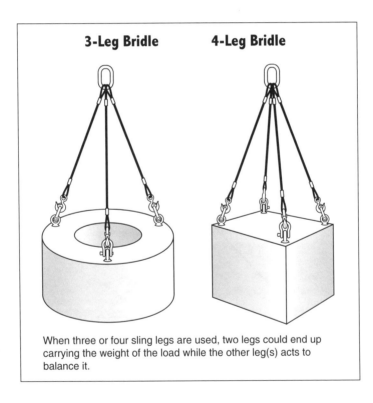

When three or four sling legs are used, two legs could end up carrying the weight of the load while the other leg(s) acts to balance it.

SLINGS • Hitches

Basket Hitches

A basket hitch is configured by wrapping or passing a sling around a load and attaching the eyes to a lifting device such as a hook. Because the load can shift or even fall out of the sling, a single basket hitch must not be used to lift loads that are difficult to balance.

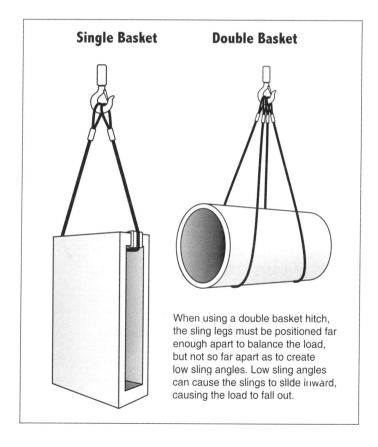

When using a double basket hitch, the sling legs must be positioned far enough apart to balance the load, but not so far apart as to create low sling angles. Low sling angles can cause the slings to slide inward, causing the load to fall out.

SLINGS • Hitches

Double Wrap Basket Hitch

A double wrap basket hitch is the same as a basket hitch with an additional wrap that goes completely around the load. Since this hitch makes full contact around the load, it is ideal for lifting loose material. Additionally, the gripping effect helps prevent the slings from sliding inward.

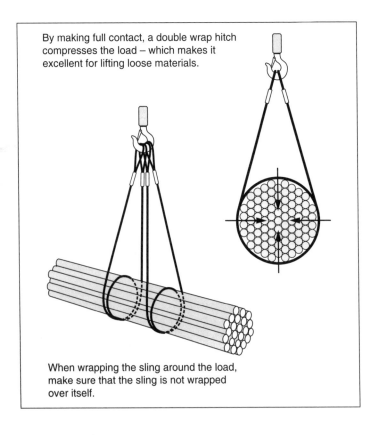

By making full contact, a double wrap hitch compresses the load – which makes it excellent for lifting loose materials.

When wrapping the sling around the load, make sure that the sling is not wrapped over itself.

SLINGS • Hitches

Choker Hitches

Rigging a choker hitch is accomplished by passing a sling around the load and through one eye or end fitting where it is then attached to a lifting hook. This type hitch is among the most commonly used because of its gripping effect on the load. When using this hitch, make sure the choke point is on the body of the sling and not on a splice or fitting.

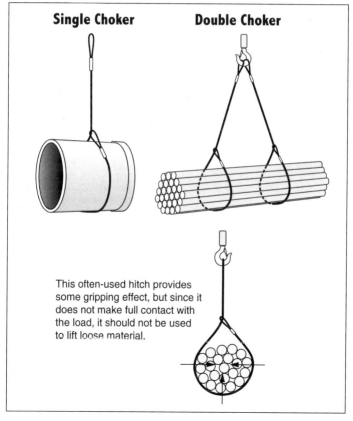

Single Choker

Double Choker

This often-used hitch provides some gripping effect, but since it does not make full contact with the load, it should not be used to lift loose material.

SLINGS • Hitches

Double Wrap Choker

A double wrap choker hitch is constructed by passing the sling twice around the load and then through one eye or fitting. This hitch is the same as a choker, with an additional wrap placed around the load.

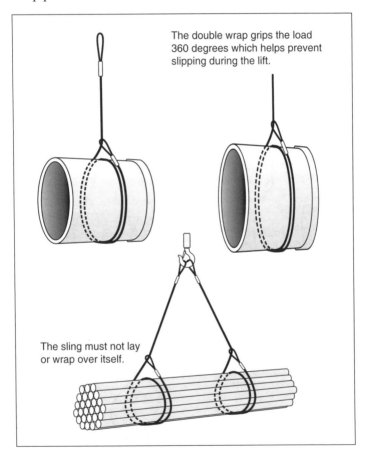

The double wrap grips the load 360 degrees which helps prevent slipping during the lift.

The sling must not lay or wrap over itself.

SLINGS • Hitches

Endless or Grommet Slings

An endless or grommet sling is made by joining the ends of the sling material to form a loop. If the material is wire rope, the ends may be hand tucked or joined by metal fittings; synthetic slings can be joined by stitching. As illustrated, these slings can be configured into a variety of choker and basket hitches.

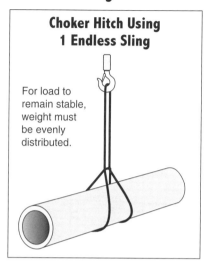

Choker Hitch Using 1 Endless Sling

For load to remain stable, weight must be evenly distributed.

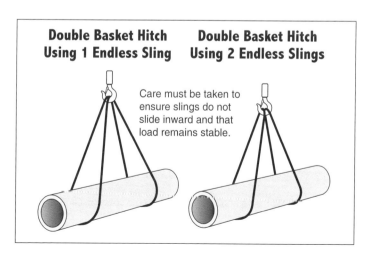

Double Basket Hitch Using 1 Endless Sling

Care must be taken to ensure slings do not slide inward and that load remains stable.

Double Basket Hitch Using 2 Endless Slings

SLINGS • Sling Angles

It is very important for slings to remain within the rated capacities that are listed in capacity tables.

The *rated capacity* of the slings used to lift a load largely depends on the angles that are formed between the sling legs and horizontal plane. As the sling angles decrease, the loading or tension on the slings increases.

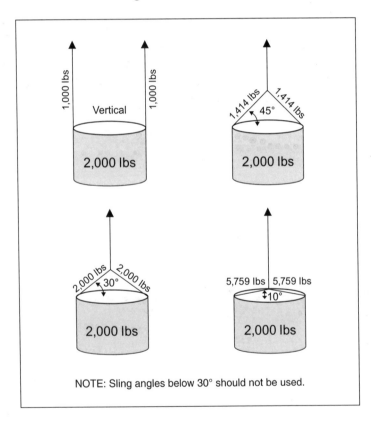

NOTE: Sling angles below 30° should not be used.

SLINGS • Sling Angles

The chart below illustrates how the tension or loading increases as sling angles decrease, especially the rapid increase in tension that occurs when slings are used below 30 degrees which is not recommended unless approved by the manufacturer or qualified person.

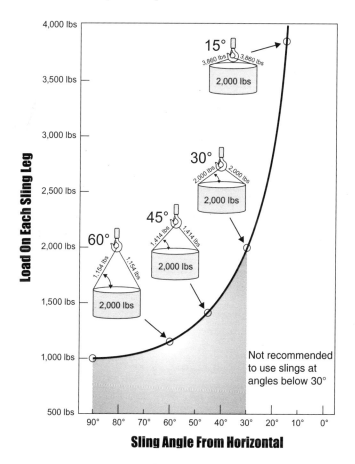

SLINGS • Determining Sling Loading

Using Capacity Tables

Example (2 Legs):

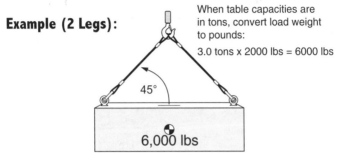

When table capacities are in tons, convert load weight to pounds:

3.0 tons x 2000 lbs = 6000 lbs

Steps:

1. Determine sling angles. (45°)
2. Go to 2-leg bridle capacity column at 45°.
3. Select 2-leg bridle with capacity equal to or greater than load to be lifted. (1/2 inch rope diameter at 7,200 lb)

Wire Rope Slings • 6 X 19 or 6 X 36 • EIPS • IWRC • MS • Rated Capacity in Pounds

Rope Diameter (Inches)	1 LEG Vertical	1 LEG Choker	Vertical or 2-Leg	BASKET AND 2 LEG BRIDLE 60 degree	45 degree	30 degree
3/8	2800	2200	5800	5000	4000	2800
7/16	3800	2800	7800	6800	5400	3800
1/2	5000	3800	10200	8800	7200	5000
9/16	6400	4800	12800	11000	9000	6400
5/8	7800	5800	15600	13600	11000	7800
3/4	11200	8200	22000	19400	15800	11200
7/8	15200	11200	30000	26000	22000	15200
1	19600	14400	40000	34000	28000	19600
1 1/8	24000	18200	48000	42000	34000	24000
1 1/4	30000	22000	60000	52000	42000	30000
1 3/8	36000	26000	72000	62000	50000	36000
1 1/2	42000	32000	84000	74000	60000	42000
1 5/8	48000	36000	98000	84000	70000	48000
1 3/4	56000	42000	114000	98000	80000	56000

• Rated capacities basket hitch based on D/d ratio of 25.
• Rated capacities based on pin diameter no larger than natural eye width or less than the nominal sling diameter.
• Horizontal sling angles less than 30 degrees shall not be used.

• When sling angles are between degree listings, use next lower angle.

Note: The increased loading created by sling angles affects the eye bolts, shackles, hooks, etc. and must be taken into consideration.

SLINGS • Determining Sling Loading

Using Sling Angles

Example (2 Legs):

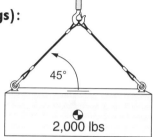

Steps:

1. Determine sling angles. (45°)
2. Select corresponding Load Angle Factor. (1.414)
3. Multiply load weight by Load Angle Factor to get total load on sling legs.
   (2,000 lb x 1.414 = 2,828 lb)
4. Divide total load by the number of sling legs.
   (2,828 lb ÷ 2 = 1,414 lb per sling leg)

Sling Angle (degrees)	Load Angle Factor
65°	1.104
60°	1.155
55°	1.221
50°	1.305
45°	1.414
40°	1.555
35°	1.742
30°	2.000

5. Select slings from the single vertical leg column within the sling capacity table.

- When sling angles are between those listed in chart, use the next lower sling angle and corresponding load angle factor.
- When the load is **not** distributed uniformly (equally) on sling legs, the tension on each leg must be calculated individually by a qualified person.

Note: The increased loading created by sling angles affects the eye bolts, shackles, hooks, etc. and must be taken into consideration.

SLINGS • Determining Sling Loading

Using Measurements

Example (2 Legs):

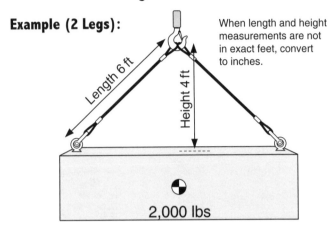

When length and height measurements are not in exact feet, convert to inches.

Steps:

1. Divide length by height to get Load Angle Factor.
   (6 ft ÷ 4 ft = 1.5)

2. Multiply Load Angle Factor by load weight to get total load on sling legs.
   (1.5 x 2,000 lb = 3,000 lb)

3. Divide total load by number of sling legs to get load on each sling leg.
   (3,000 lb ÷ 2 = 1,500 lb per sling leg)

4. Select slings from the single vertical leg column within the sling capacity table.

   - When the load is **not** distributed uniformly (equally) on sling legs, the tension on each leg must be calculated individually by a qualified person.

   Note: The increased loading created by sling angles affects the eye bolts, shackles, hooks, etc. and must be taken into consideration.

SLINGS • Determining Sling Loading

Using Measurements

Example (3 Legs):

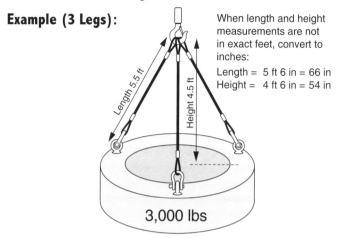

When length and height measurements are not in exact feet, convert to inches:

Length = 5 ft 6 in = 66 in
Height = 4 ft 6 in = 54 in

Steps:

1. Divide length by height to get Load Angle Factor.
   (66 in ÷ 54 in = 1.2)

2. Multiply Load Angle Factor by load weight to get total load on sling legs.
   (1.2 x 3,000 lb = 3,600 lb)

3. Divide total load by number of sling legs to get load on each sling leg.
   (3,600 lb ÷ 3 = 1,200 lb per sling leg)

4. Select slings from the single vertical leg column within the sling capacity table.

   - When the load is **not** distributed uniformly (equally) on sling legs, the tension on each leg must be calculated individually by a qualified person.

   Note: The increased loading created by sling angles affects the eye bolts, shackles, hooks, etc. and must be taken into consideration.

SLINGS • Determining Sling Loading

Using Measurements

Example (2 Leg Choker):

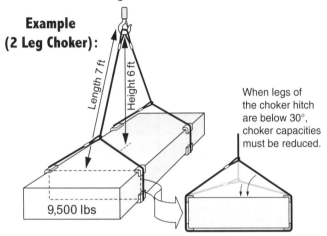

When legs of the choker hitch are below 30°, choker capacities must be reduced.

Steps:

1. Divide length by height to get Load Angle Factor.
   (7 ft ÷ 6 ft = 1.2)

2. Multiply Load Angle Factor by load weight to get total load on sling legs.
   (1.2 x 9,500 lb = 11,400 lb)

3. Divide total load by number of sling legs to get load on each sling leg.
   (11,400 lb ÷ 2 = 5,700 lb per sling leg)

4. Select slings from the single leg choker column within the sling capacity table.

   - When the load is **not** distributed uniformly (equally) on sling legs, the tension on each leg must be calculated individually by a qualified person.

   Note: The increased loading created by sling angles affects the eye bolts, shackles, hooks, etc. and must be taken into consideration.

SLINGS • Determining Sling Loading

Using Measurements: Unequal Legs

When slings legs are equal lengths, the loading or tension on each leg will be the same. However, when the center of gravity of the load is offset and the sling legs are different lengths, the loading on each leg will be different and must be calculated individually.

Example (2 Legs):

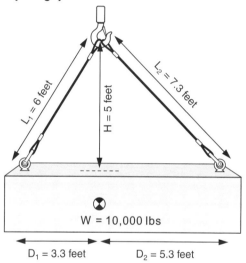

To determine the tension on Leg 1 and Leg 2:

$$\text{Leg 1} = \frac{W \times D_2 \times L_1}{H \times (D_1 + D_2)} = \frac{10{,}000 \text{ lb} \times 5.3 \text{ ft} \times 6 \text{ ft}}{5 \text{ ft} \times (3.3 \text{ ft} + 5.3 \text{ ft})} = 7{,}395 \text{ lb}$$

$$\text{Leg 2} = \frac{W \times D_1 \times L_2}{H \times (D_1 + D_2)} = \frac{10{,}000 \text{ lb} \times 3.3 \text{ ft} \times 7.3 \text{ ft}}{5 \text{ ft} \times (3.3 \text{ ft} + 5.3 \text{ ft})} = 5{,}602 \text{ lb}$$

Note: The increased loading created by sling angles affects the eye bolts, shackles, hooks, etc. and must be taken into consideration.

SLINGS • Determining Sling Loading

Using Measurements: Unequal Legs and Heights

Example (2 Legs):

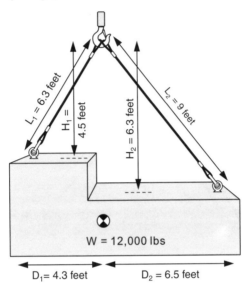

To determine the tension on Leg 1 and Leg 2:

$$\text{Leg 1} = \frac{W \times D_2 \times L_1}{(D_2 \times H_1) + (D_1 \times H_2)} = \frac{12{,}000 \text{ lb} \times 6.5 \text{ ft} \times 6.3 \text{ ft}}{(6.5 \text{ ft} \times 4.5 \text{ ft}) + (4.3 \text{ ft} \times 6.3 \text{ ft})} = 8{,}722 \text{ lb}$$

$$\text{Leg 2} = \frac{W \times D_1 \times L_2}{(D_2 \times H_1) + (D_1 \times H_2)} = \frac{12{,}000 \text{ lb} \times 4.3 \text{ ft} \times 9 \text{ ft}}{(6.5 \text{ ft} \times 4.5 \text{ ft}) + (4.3 \text{ ft} \times 6.3 \text{ ft})} = 8{,}243 \text{ lb}$$

Note: The increased loading created by sling angles affects the eye bolts, shackles, hooks, etc. and must be taken into consideration.

SLINGS • Determining Sling Loading

Using Measurements: Drifting Load

Since the structures that are normally used as attached points are not usually designed for side pulls, they could be damaged from the tension created. Before using this method for lifting and moving loads, a qualified engineer should be consulted to ensure the structure will withstand the imposed forces.

Tension will change as the load is drifted from one place to another. To avoid overloading the chainfall and/or other lifting devices, their selection must be based on the highest tension created at any one point.

Example (2 Legs):

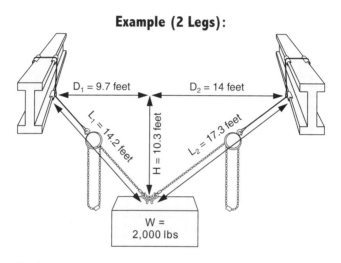

To determine the tension on Leg 1 and Leg 2:

$$\text{Leg 1} = \frac{W \times D_2 \times L_1}{H \times (D_1 + D_2)} = \frac{2{,}000 \text{ lb} \times 14 \text{ ft} \times 14.2 \text{ ft}}{10.3 \text{ ft} \times (9.7 \text{ ft} + 14 \text{ ft})} = 1{,}629 \text{ lb}$$

$$\text{Leg 2} = \frac{W \times D_1 \times L_2}{H \times (D_1 + D_2)} = \frac{2{,}000 \text{ lb} \times 9.7 \text{ ft} \times 17.3 \text{ ft}}{10.3 \text{ ft} \times (9.7 \text{ ft} + 14 \text{ ft})} = 1{,}375 \text{ lb}$$

Note: The increased loading created by sling angles affects the eye bolts, shackles, hooks, etc. and must be taken into consideration.

SLINGS • Overlooked Sling Loading

Often overlooked are certain hitch configurations that can have angles which cause extra tension or loading to be created. Illustrated are examples of two hitches commonly used in the field where this occurs. Avoid these angles when possible. Otherwise, they must be considered.

Capacities for slings configured in vertical basket hitches are based on the legs being 5° or less from the vertical.

When the legs are greater than 5°, additional sling loading must be taken into consideration.

For choker hitches, the rated capacities are based on the horizontal angles of the legs being 30° or greater.

To better grip the load, a common practice in the field is to beat the choker legs down. When the legs are forced down, the low angles increase tension and sling capacity is reduced. Avoid using leg angles below 30° from the horizontal.

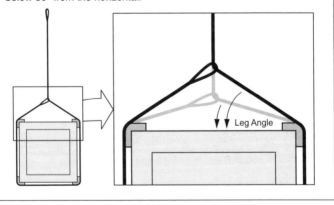

SLINGS • Overlooked Sling Loading

Choker Hitches

When a load is hanging free, the normal *angle of choke* is approximately 135°. When this angle is 120° or less, a reduction in the sling capacity must be made. This applies to all type slings – see chart below.

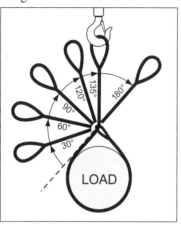

Choker Hitch Capacity Adjustment

Angle of Choke (degrees)	Rated Capacity (percent)
over 120	100
90-120	87
60-89	74
30-59	62
0-29	49

This applies to all type slings

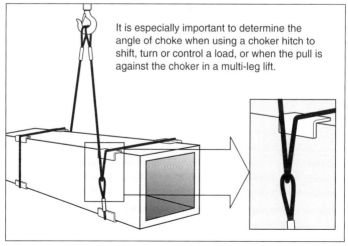

It is especially important to determine the angle of choke when using a choker hitch to shift, turn or control a load, or when the pull is against the choker in a multi-leg lift.

SLINGS • Wire Rope Slings

Rated capacities for wire rope slings used in a basket hitch are based on a round contact surface of 25 times sling diameter; 15 times for hand-tucked slings. When using softeners, sling capacity will be reduced depending on the radius created by the contact surface of the softeners (see D/d ratio chart on next page). As the radius decreases, bending of the sling increases, resulting in loss of strength.

Slings with fiber cores must not be exposed to temperatures above 180°F and those with IWRC should not be exposed to temperatures above 400°F or below -40°F without consulting the sling manufacturer.

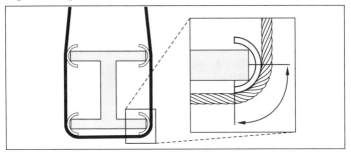

Sling Eyes

The eye of a wire rope sling should never be used or forced over a hook, pin or object where the body diameter (D) is greater than one-half the length of the eye. The angles of the eye that form the triangle create tension the same way sling legs do when used at angles. Because of the severe bending that can occur, the hook or pin must also be no smaller than the nominal diameter of the sling body.

SLINGS • Wire Rope Slings

D/d Ratio

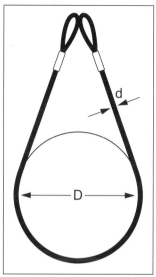

When a wire rope sling is used in a basket hitch, the diameter of the load where the sling contacts the load can reduce sling capacity. The method used to determine the loss of strength or efficiency is referred to as D/d ratio. "D" refers to the diameter of the object; "d" refers to the diameter of the wire rope sling. For example, when the diameter of the object is 25 times the diameter of the sling, the D/d ratio is 25/1.

To ensure that slings are not overloaded, utilize the D/d ratios in the chart below.

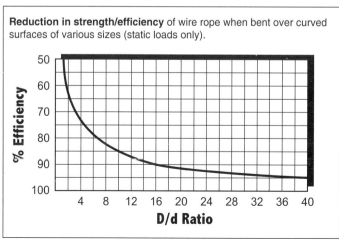

Reduction in strength/efficiency of wire rope when bent over curved surfaces of various sizes (static loads only).

SLINGS • Wire Rope Slings

Hand-Spliced Slings

Wire rope slings constructed with a *hand-tucked splice* should *not* be used in a single vertical lift. The rotation of the load could cause the sling to unlay and result in loss of load.

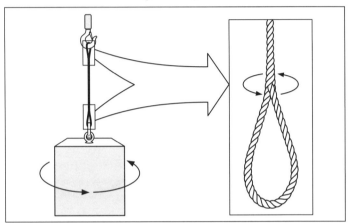

Wire rope clips must not be used to fabricate slings unless the application prevents the use of prefabricated slings or where the specific application is designed by a qualified person. However, they must never be used as a choker hitch and the use of clips must not be used to shorten or lengthen the sling.

SLINGS • Wire Rope Slings

Inspection

Wire rope slings must be inspected by a competent person each day before being used and regularly during use. Slings must be removed from service when any of the following conditions exist:

- Hooks and latches deformed or damaged (see page 99).
- No identification stating manufacturer, rated load, number of legs, and diameter.
- For strand laid and single part slings:

    10 randomly distributed broken wires in one rope lay,
    5 broken wires in one strand in one rope lay, or
    2 broken wire at end attachment.

- Severe localized abrasion or scraping.
- Kinking, crushing, unstranding, birdcaging, or any other damage resulting in distortion of the rope structure.
- Evidence of heat damage.
- End attachments that are cracked, deformed, or worn.
- Corrosion of the rope or end attachments.

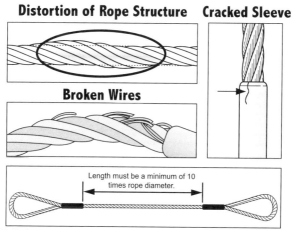

Distortion of Rope Structure   **Cracked Sleeve**

Broken Wires

Length must be a minimum of 10 times rope diameter.

NOTE: Refer also to pages 13-15 on wire rope inspection.

WIRE ROPE SLINGS • 1-Part Hand Tucked Splice

6 x 19, 6 x 37 • EIPS • FC & IWRC • Rated Capacity in Tons

Rope Diameter (inches)	1 LEG Vertical	1 LEG Choker	BASKET & 2 LEG BRIDLE Vertical Basket or 2-Leg	60 degree	45 degree	30 degree
3/8	1.2	0.94	2.4	2.0	1.7	1.2
7/16	1.6	1.3	3.2	2.7	2.2	1.6
1/2	2.0	1.6	4.0	3.5	2.9	2.0
9/16	2.5	2.1	5.0	4.4	3.6	2.5
5/8	3.1	2.6	6.2	5.3	4.4	3.1
3/4	4.3	3.7	8.6	7.4	6.1	4.3
7/8	5.7	5.0	11	9.8	8.0	5.7
1	7.4	6.4	15	13	10	7.4
1 1/8	9.3	8.1	19	16	13	9.3
1 1/4	11	9.9	23	20	16	11
1 3/8	14	12	27	24	19	14
1 1/2	16	14	32	28	23	16
1 5/8	19	16	38	33	27	19
1 3/4	22	19	44	38	31	22
1 7/8	25	22	50	43	35	25
2	28	25	56	49	40	28
2 1/8	32	28	63	55	45	32
2 1/4	35	31	70	61	50	35
2 3/8	39	34	78	68	55	39
2 1/2	43	38	86	74	61	43

- Rated capacities based on design factor of 5.
- Rated capacities for basket hitches based on D/d ratio of 15.
- Rated capacities based on pin diameter no larger than natural eye width or less than the nominal sling diameter.
- Horizontal sling angles less than 30 degrees shall not be used.

WIRE ROPE SLINGS • 1-Part Mechanical Splice

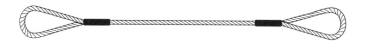

6 x 19 and 6 x 37 • EIPS • IWRC • Rated Capacity in Tons

Rope Diameter (inches)	1 LEG Vertical	1 LEG Choker	BASKET & 2 LEG BRIDLE Vertical Basket or 2-Leg	60 degree	45 degree	30 degree
3/8	1.4	1.1	2.9	2.5	2.0	1.4
7/16	1.9	1.4	3.9	3.4	2.7	1.9
1/2	2.5	1.9	5.1	4.4	3.6	2.5
9/16	3.2	2.4	6.4	5.5	4.5	3.2
5/8	3.9	2.9	7.8	6.8	5.5	3.9
3/4	5.6	4.1	11	9.7	7.9	5.6
7/8	7.6	5.6	15	13	11	7.6
1	9.8	7.2	20	17	14	9.8
1 1/8	12	9.1	24	21	17	12
1 1/4	15	11	30	26	21	15
1 3/8	18	13	36	31	25	18
1 1/2	21	16	42	37	30	21
1 5/8	24	18	49	42	35	24
1 3/4	28	21	57	49	40	28
1 7/8	32	24	64	56	46	32
2	37	28	73	63	52	37
2 1/8	40	31	80	69	56	40
2 1/4	44	35	89	77	63	44
2 3/8	49	38	99	85	70	49
2 1/2	54	42	109	94	77	54
2 5/8	60	46	119	103	84	60
2 3/4	65	51	130	113	92	65

- Rated capacities based on design factor of 5.
- Rated capacities for basket hitches based on D/d ratio of 25.
- Rated capacities based on pin diameter no larger than natural eye width or less than the nominal sling diameter.
- Horizontal sling angles less than 30 degrees shall not be used.

WIRE ROPE SLINGS • 1-Part Mechanical Splice

6 x 19 and 6 x 37 • EIPS • IWRC • Rated Capacity in Tons

Rope Diameter (inches)	3 LEG BRIDLE				4 LEG BRIDLE			
	Vertical	60 degree	45 degree	30 degree	Vertical	60 degree	45 degree	30 degree
3/8	4.3	3.7	3.0	2.2	5.7	5.0	4.1	2.9
7/16	5.8	5.0	4.1	2.9	7.8	6.7	5.5	3.9
1/2	7.6	6.6	5.4	3.8	10	8.8	7.1	5.1
9/16	9.6	8.3	6.8	4.8	13	11	9.0	6.4
5/8	12	10	8.3	5.9	16	14	11	7.8
3/4	17	15	12	8.4	22	19	16	11
7/8	23	20	16	11	30	26	21	15
1	29	26	21	15	39	34	28	20
1 1/8	36	31	26	18	48	42	34	24
1 1/4	44	38	31	22	59	51	42	30
1 3/8	53	46	38	27	71	62	50	36
1 1/2	63	55	45	32	84	73	60	42
1 5/8	73	63	52	37	98	85	69	49
1 3/4	85	74	60	42	113	98	80	57
1 7/8	97	84	68	48	129	112	91	64
2	110	95	78	55	147	127	104	73
2 1/8	119	103	84	60	159	138	112	80
2 1/4	133	116	94	67	178	154	126	89
2 3/8	148	128	105	74	197	171	139	99
2 1/2	163	141	115	82	217	188	154	109
2 5/8	179	155	126	89	238	206	168	119
2 3/4	195	169	138	97	260	225	184	130

- Rated capacities based on design factor of 5.
- Rated capacities for basket hitches based on D/d ratio of 25.
- Rated capacities based on pin diameter no larger than natural eye width or less than the nominal sling diameter.
- Horizontal sling angles less than 30 degrees shall not be used.

WIRE ROPE SLINGS • 1-Part Mechanical Splice

6 x 19 and 6 x 37 • EIPS • IWRC • Rated Capacity in Tons
2 LEG CHOKER

Rope Diameter (inches)	Vertical	60 degree	45 degree	30 degree
3/8	2.1	1.8	1.5	1.1
7/16	2.9	2.5	2.0	1.4
1/2	3.7	3.2	2.6	1.9
9/16	4.7	4.1	3.3	2.4
5/8	5.8	5.0	4.1	2.9
3/4	8.2	7.1	5.8	4.1
7/8	11	9.7	7.9	5.6
1	14	13	10	7.2
1 1/8	18	16	13	9.1
1 1/4	22	19	16	11
1 3/8	27	23	19	13
1 1/2	32	28	23	16
1 5/8	37	32	26	18
1 3/4	43	37	30	21
1 7/8	49	42	34	24
2	55	48	39	28
2 1/8	62	54	44	31
2 1/4	69	60	49	35
2 3/8	77	66	54	38
2 1/2	85	73	60	42
2 5/8	93	80	66	46
2 3/4	101	88	71	51

- Rated capacities based on design factor of 5.
- Rated capacities for basket hitches based on D/d ratio of 25.
- Rated capacities based on pin diameter no larger than natural eye width or less than the nominal sling diameter.
- Horizontal sling angles less than 30 degrees shall not be used.

WIRE ROPE SLINGS • 8-Part and 6-Part

6 x 19 and 6 x 37 • EIPS • IWRC • Rated Capacity in Tons
SINGLE LEG

Component Rope Diameter (inches)	Vertical		Choker		Vertical Basket	
	8 Part	6 Part	8 Part	6 Part	8 Part*	6 Part*
1/4	3.8	2.9	3.3	2.5	7.6	5.7
5/16	5.9	4.4	5.2	3.9	12	8.8
3/8	8.5	6.3	7.4	5.5	17	13
7/16	11	8.6	10	7.5	23	17
1/2	15	11	13	9	30	22
9/16	19	14	16	12	38	28
5/8	23	17	20	15	46	35
3/4	33	25	29	22	66	49
7/8	45	33	39	29	89	67
1	58	43	51	38	116	87
1 1/8	73	55	64	48	146	109
1 1/4	89	67	78	59	179	134
1 3/8	108	81	94	71	215	161
1 1/2	127	96	111	84	255	191
1 5/8	148	111	130	97	296	222
1 3/4	171	128	150	112	343	257
1 7/8	195	146	171	128	390	292
2	222	166	194	145	443	332

- These values are based on slings being vertical. If they are not vertical, the rated capacity shall be reduced. If two or more slings are used, the lowest sling angle from the horizontal shall be considered:
- Rated capacities for multiple sling legs = rated capacity of single leg sling x number of sling legs x sine of the lowest sling angle from the horizontal.

* These values only apply when the D/d ratio is 25 or greater.

SLINGS • Metal Mesh Slings

Metal mesh slings are useful in applications where abrasion, heat and corrosion are factors. Their smooth flat surface has a gripping effect on the load without stretching and can conform to irregular shapes. To prevent the sling or load from being damaged, they can be coated with rubber or plastic. When used below -20°F or above 550°F, the rated capacity must be reduced. Elastometer coated metal mesh slings can only be used in a temperature range from 0°F to 200°F.

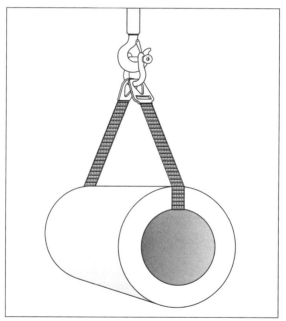

Metal mesh slings are available in three strengths, commonly referred to as gauges: 14 gauge (light duty), 12 gauge (medium duty), and 10 gauge (heavy duty).

Any repairs to metal mesh slings should be made by the sling manufacturer.

SLINGS • Metal Mesh Slings

Inspection

Metal mesh slings must be inspected by a competent person each day before being used and regularly during use. Slings must be removed from service when any of the following conditions exist:

- A broken weld or a broken brazed joint along the sling edge.
- A broken wire in any part of the mesh.
- Reduction in wire diameter of 25% due to abrasion or 15% due to corrosion.
- Lack of flexibility due to distortion of the mesh; locked spirals.
- Distortion of the choker fitting so the depth of the slot is increased by more than 10%.
- Distortion of either end fitting so the width of the eye opening is decreased by more than 10%.
- A 15% reduction of the original cross-sectional area of metal at any point around the hook opening or end fitting.
- Visible distortion of either end fitting out of its plane.
- Any fittings that are pitted, corroded, cracked, bent, twisted, gouged, broken or has sharp edges.
- Missing or illegible sling identification.

Broken Wires

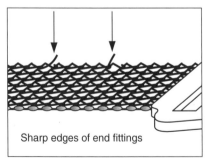

Sharp edges of end fittings

Bent End Fitting

SLINGS • Metal Mesh Slings

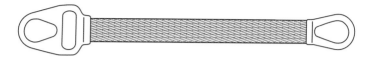

Rated Capacities in Pounds

Sling Width (inches)	Vertical or Choker	Vertical Basket	60 degree	45 degree	30 degree
LIGHT DUTY — 14 GAUGE					
2	900	1800	1560	1270	900
3	1400	2800	2420	1980	1400
4	2000	4000	4150	3390	2000
6	3000	6000	5190	4240	3000
8	4000	8000	6920	5650	4000
10	5000	10000	8660	7070	5000
12	6000	12000	10390	8480	6000
14	7000	14000	12120	9890	7000
16	8000	16000	13850	11310	8000
18	9000	18000	15580	12720	9000
20	10000	20000	17320	14140	10000
MEDIUM DUTY — 12 GAUGE					
2	1450	2900	2510	2050	1450
3	2170	4350	3770	3070	2170
4	2900	5800	5020	4100	2900
6	4800	9600	8310	6780	4800
8	6400	12800	11080	9050	6400
10	8000	16000	13850	11310	8000
12	9600	19200	16620	13570	9600
14	11200	22400	19400	15830	11200
16	12800	25600	22170	18100	12800
18	13500	27000	23380	19090	13500
20	15000	30000	25980	21210	15000

• All angles shown are measured from the horizontal.

SLINGS • Metal Mesh Slings

Rated Capacities in Pounds

Sling Width (inches)	Vertical or Choker	Vertical Basket	60 degree	45 degree	30 degree
HEAVY DUTY — 10 GAUGE					
2	1600	3200	2770	2260	1600
3	3000	6000	5200	4240	3000
4	4400	8800	7620	6220	4400
6	6600	13200	11430	9330	6600
8	8800	17600	15240	12440	8800
10	11000	22000	19050	15550	11000
12	13200	26400	22860	18660	13200
14	15400	30800	26670	21770	15400
16	17600	35200	30480	24880	17600
18	19800	39600	34290	28000	19800
20	22000	44000	38100	31100	22000

• All angles shown are measured from the horizontal.

SLINGS • Chain Slings

Because of the rugged nature of chain slings, they are more useful in certain environments and applications than other types of slings. They are better suited to resist abrasion, corrosion, heat, and sharp edges.

It is important to remember that only chain slings made from alloy steel be used for overhead lifting. They are commonly identified by the numbers 8, 80 or 800 for grade 80 and 10, 100 or 1000 for grade 100.

There are other grades of proof tested steel chain, such as Proof Coil (grade 28) and Hi-Test (grade 43). However, these grades are not recommended for overhead lifting.

The hook must not be inserted into one of the chain links, and like other type slings, sharp corners of the load must be padded.

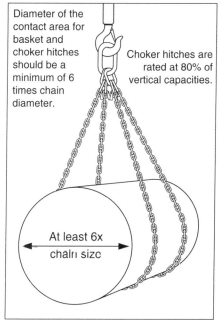

Diameter of the contact area for basket and choker hitches should be a minimum of 6 times chain diameter.

Choker hitches are rated at 80% of vertical capacities.

At least 6x chain size

SLINGS • Chain Slings

Chain slings must never be used when twisted, knotted, or whenever the links bind and do not move freely.

Never use makeshift hooks, links or nuts, and bolts, etc. to repair chain or change length. Any repairs should be made by the chain manufacturer.

When chain slings are used in temperatures of 400°F and above, rated capacities must be reduced in accordance with the manufacturer's recommendations. The chain manufacturer should also be consulted whenever chain slings are to be used in temperatures of −40°F or below.

Alloy Chain: Effect of Elevated Temperature on Rated Load Limit				
	Grade 80		Grade 100	
Temperature degrees Fahrenheit	Temporary reduction of rated load while at temperature	Permanent reduction of rated load after exposure to temperature	Temporary reduction of rated load while at temperature	Permanent reduction of rated load after exposure to temperature
below 400	none	none	none	none
400 – 499	10%	none	15%	none
500 – 599	15%	none	25%	5%
600 – 699	20%	5%	30%	15%
700 – 799	30%	10%	40%	20%
800 – 899	40%	15%	50%	25%
900 – 999	50%	20%	60%	30%
1000 and above	remove from service	remove from service	remove from service	remove from service

SLINGS • Chain Slings

Inspection

Chain slings are required to be inspected by a competent person each day before being used, during use, and periodically with a record kept. Inspection includes the chain itself and all attachments for wear, nicks, cracks, breaks, gouges, stretching, bending, twisting, weld spatter, corrosion, discoloration from excessive temperature, and damage to hooks and hook latches.

- Chain links and attachments must hinge freely.

- Worn links shall not exceed values given in the table below.

- Sharp transverse nicks and gouges should be rounded out by grinding. The depth of grinding should not exceed values in the table below.

- Repairs must only be made by the chain sling manufacturer or qualified person, and be permanently marked to identify the repairing agency.

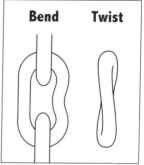

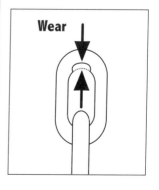

Minimum Allowable Thickness at Any Point of Link

Nominal chain or coupling link size		Minimum allowable thickness at any point on link	
(inches)	(mm)	(inches)	(mm)
7/32	5.5	0.189	4.80
9/32	7	0.239	6.07
5/16	8	0.273	6.93
3/8	10	0.342	8.69
1/2	13	0.443	11.26
5/8	16	0.546	13.87
3/4	20	0.687	17.15
7/8	22	0.750	19.05
1	26	0.887	22.53
1 1/4	32	1.091	27.71

NOTE: For other sizes, consult the chain or sling manufacturer.

SLINGS • Chain Slings

Inspection

To determine if the sling has been stretched, measure the chain as shown and compare the measurement to the **length** or **reach** listed on the identification tag.

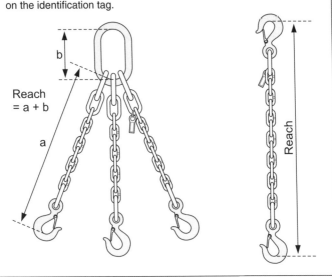

Each alloy steel chain sling is required to have an affixed, durable identification tag.

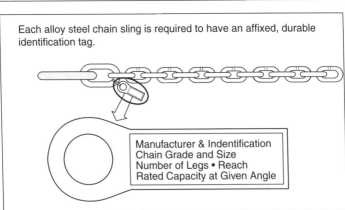

Manufacturer & Indentification
Chain Grade and Size
Number of Legs • Reach
Rated Capacity at Given Angle

SLINGS • Alloy Steel Chain, Grade 80

Single Leg & 2 Leg

Working Load Limit, in Pounds • 4 to 1 Design Factor

Chain Size		SINGLE LEG	2 LEG		
		90 deg vertical	60 deg	45 deg	30 deg
inches	mm				
7/32	5.5	2100	3600	3000	2100
9/32	7	3500	6100	4900	3500
5/16	8	4500	7800	6400	4500
3/8	10	7100	12300	10000	7100
1/2	13	12000	20800	17000	12000
5/8	16	18100	31300	25600	18100
3/4	20	28300	49000	40000	28300
7/8	22	34200	59200	48400	34200
1	26	47700	82600	67400	47700
1 1/4	32	72300	125200	102200	72300

- Rated capacities for slings used in a choker hitch shall be a maximum of 80% of the rated capacities for single and multiple leg slings provided that the angle of choke is greater than 120 degrees.

- The horizontal angle is the angle formed between the inclined leg and the horizontal plane of the load.

SLINGS • Alloy Steel Chain, Grade 80

3 Leg & 4 Leg

Working Load Limit, in Pounds • 4 to 1 Design Factor

Chain Size		3 LEG & 4 LEG		
		60 deg	45 deg	30 deg
inches	mm			
7/32	5.5	5500	4400	3200
9/32	7	9100	7400	5200
5/16	8	11700	9500	6800
3/8	10	18400	15100	10600
1/2	13	31200	25500	18000
5/8	16	47000	38400	27100
3/4	20	73500	60000	42400
7/8	22	88900	72500	51300
1	26	123900	101200	71500
1 1/4	32	187800	153400	108400

- Chain slings made with grades of steel chain other than Grades 80 and 100 alloy steel are not recommended for overhead lifting.
- Rating of multileg slings adjusted for angle of loading between the inclined leg and the horizontal plane of the load.
- Quadruple sling rating is same as triple sling because normal lifting practice may not distribute load uniformly on all four legs.

SLINGS • Alloy Steel Chain, Grade 100

Single Leg & 2 Leg

Working Load Limit, in Pounds • 4 to 1 Design Factor

Chain Size		SINGLE LEG	2 LEG		
inches	mm	90 deg vertical	60 deg	45 deg	30 deg
7/32	5.5	2700	4700	3800	2700
9/32	7	4300	7500	6100	4300
5/16	8	5700	9900	8100	5700
3/8	10	8800	15200	12400	8800
1/2	13	15000	26000	21200	15000
5/8	16	22600	39100	32000	22600
3/4	20	35300	61100	49900	35300
7/8	22	42700	74000	60400	42700

- Rated capacities for slings used in a choker hitch shall be a maximum of 80% of the rated capacities for single and multiple leg slings provided that the angle of choke is greater than 120 degrees.
- The horizontal angle is the angle formed between the inclined leg and the horizontal plane of the load.

SLINGS • Alloy Steel Chain, Grade 100

3 Leg & 4 Leg

Working Load Limit, in Pounds • 4 to 1 Design Factor

Chain Size		3 LEG & 4 LEG		
inches	mm	60 deg	45 deg	30 deg
7/32	5.5	7000	5000	4000
9/32	7	11200	9100	6400
5/16	8	14800	12100	8500
3/8	10	22800	18600	13200
1/2	13	39000	31800	22500
5/8	16	58700	47900	33900
3/4	20	91700	74900	53000
7/8	22	110900	90600	64000

- Chain slings made with grades of steel chain other than Grades 80 and 100 alloy steel are not recommended for overhead lifting.
- Rating of multileg slings adjusted for angle of loading between the inclined leg and the horizontal plane of the load.
- Quadruple sling rating is same as triple sling because normal lifting practice may not distribute load uniformly on all four legs.

SLINGS • Synthetic Web Slings

Light weight and ease of handling make synthetic web slings popular. However, this can cause them to be used in applications requiring a more rugged and durable sling such as wire rope, mesh, or chain.

Web slings are good for use on machined, polished, and painted surfaces. Being non-sparking, they also work well in combustible environments. However, since they can be easily cut or torn, web slings should not be used against sharp edges. Synthetic web slings should not be used in excess of 194 degrees or below -45 degrees Fahrenheit.

As illustrated, synthetic web slings are available in a variety of types, configurations and classes.

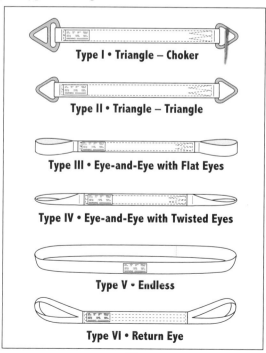

Type I • Triangle – Choker

Type II • Triangle – Triangle

Type III • Eye-and-Eye with Flat Eyes

Type IV • Eye-and-Eye with Twisted Eyes

Type V • Endless

Type VI • Return Eye

SLINGS • Synthetic Web Slings

Synthetic web slings must have no knots, and because of their makeup they must not be dragged across floors or other objects, and must never be pulled from under loads.

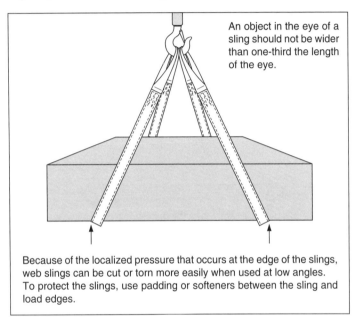

An object in the eye of a sling should not be wider than one-third the length of the eye.

Because of the localized pressure that occurs at the edge of the slings, web slings can be cut or torn more easily when used at low angles. To protect the slings, use padding or softeners between the sling and load edges.

To help protect against abrasion and other damage, synthetic web slings are available with edge guards that are sewn in or slide along the sling.

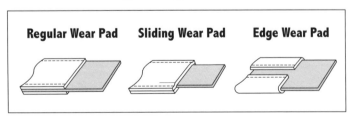

SLINGS • Synthetic Web Slings

Inspection

Synthetic web slings must be inspected by a competent person each day before being used and regularly during use. Slings must be removed from service when any of the following conditions exist:

- Sling identification missing, illegible, or incomplete.
- Chemical damage including acid or caustic burns, brittle or stiff areas, and discoloration on any part of the sling.
- Melting or charring of any part of the sling surface.
- Knots, snags, holes, tears, cuts, or extensive abrasive wear.
- Broken or worn stitches.
- Corrosion, distortion, or other damage to fittings.

Some synthetic web slings are made with red yarns or threads that when visible indicate wear has exceeded the permitted limits.

Any repairs to synthetic web slings should be made by the sling manufacturer and proof tested (with a record kept) before being put back into service.

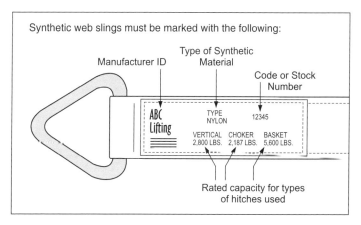

SYNTHETIC WEB SLINGS • 1-Ply

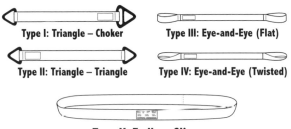

Type I: Triangle – Choker

Type II: Triangle – Triangle

Type III: Eye-and-Eye (Flat)

Type IV: Eye-and-Eye (Twisted)

Type V: Endless Slings

Rated Load for One-Ply, Class 5 Synthetic Webbing Slings

	Types I, II, III, IV			Two-Leg or Single Basket				Type V
	Sling Leg			Horizontal Angle				
	Vertical	Choker	Vertical Basket	Vertical	60 deg	45 deg	30 deg	Endless Vertical
Width, in.								
1	1100	880	2200	2200	1900	1600	1100	2200
1-1/2	1600	1280	3200	3200	2800	2300	1600	3200
1-3/4	1900	1520	3800	3800	3300	2700	1900	3800
2	2200	1760	4400	4400	3800	3100	2200	4400
3	3300	2640	6600	6600	5700	4700	3300	6600
4	4400	3520	8800	8800	7600	6200	4400	8800
5	5500	4400	11000	11000	9500	7800	5500	11000
6	6600	5280	13200	13200	11400	9300	6600	13200

NOTES:
- The rated loads are based on stuffer weave construction webbing with a minimum certified tensile strength of 6,800 lb/in. of width of the webbing.
- Rated loads for Types III and IV slings apply to both tapered and non tapered eye constructions. Rated loads for Type V slings are based on nontapered webbing.
- For Type VI slings, consult the manufacturer for rated loads.
- For choker hitch, the angle of choke shall be 120 deg or greater (see page 59 and ASME B30.9-5.5.4)

SLINGS • Synthetic Roundslings

Synthetic "roundslings" are endless and/or eye and eye slings consisting of a load-bearing synthetic yarn core which is enclosed in a protective synthetic cover. These slings can be manufactured with end fittings, center fittings, or no fittings.

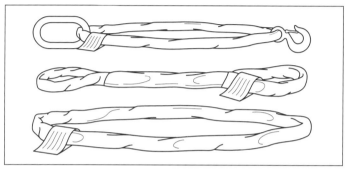

Inspection

Synthetic roundslings must be inspected by a competent person each day before use, and regularly during use. Slings must be removed from service when any of the following conditions exist:

- Missing or illegible sling identification (required: manufacturer ID, code or stock number, rated capacity for hitches, core and cover material).
- Chemical damage including acid or caustic burns, brittle or stiff areas, and discoloration on any part of the sling.
- Melting, charring, or weld spatter on any part of the fittings.
- Holes, tears, cuts, snags, broken or worn stitching, or any abrasion in the sling cover which exposes the core yarns.
- Knotting of the sling.
- Stretching, cracking, pitting, distortion, or any other damage to the fittings.
- Other visible damage that could affect sling strength.

SYNTHETIC ROUNDSLINGS • Single Leg Polyester

Endless and Eye-and-Eye • Rated Capacity in Pounds

Size	Vertical	Choker	90 degree	60 degree	45 degree	30 degree
1	2600	2100	5200	4500	3700	2600
2	5300	4200	10600	9200	7500	5300
3	8400	6700	16800	14500	11900	8400
4	10600	8500	21200	18400	15000	10600
5	13200	10600	26400	22900	18700	13200
6	16800	13400	33600	29100	23800	16800
7	21200	17000	42400	36700	30000	21200
8	25000	20000	50000	43300	35400	25000
9	31000	24800	62000	53700	43800	31000
10	40000	32000	80000	69300	56600	40000
11	53000	42400	106000	91800	74900	53000
12	66000	52800	132000	114300	93300	66000
13	90000	72000	180000	155900	127300	90000

- The size numbers in the first column of this table are those adopted by the Web
- Sling and Tiedown Association to describe certain polyester roundslings, and are included for reference only; other polyester roundslings may have different vertical rated loads.
- Color guidelines for the covers of polyester roundslings are widely used to indicate the vertical rated capacity, but this system is not followed by some manufacturers. Always select and use roundslings by the rated capacity shown on the tag, not by color.

SLINGS • Synthetic Rope Slings

Synthetic rope slings are made from a variety of synthetic fibers such as nylon, polyester or polypropylene, with nylon being the most commonly used. Those made from polyester or nylon are not to be used in temperatures above 194°F or below -40°F.

All synthetic ropes lose some strength when wet, with nylon rope losing about 10%. It is important to remember that all rope conducts electricity when wet, especially nylon which is not recommended for use around power lines or other power sources.

Synthetic rope slings must only be used to lift light loads with care taken when working around heat and sharp edges.

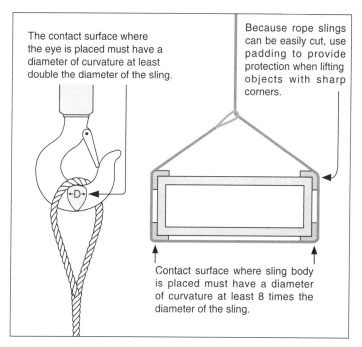

The contact surface where the eye is placed must have a diameter of curvature at least double the diameter of the sling.

Because rope slings can be easily cut, use padding to provide protection when lifting objects with sharp corners.

Contact surface where sling body is placed must have a diameter of curvature at least 8 times the diameter of the sling.

SLINGS • Synthetic Rope Slings

Inspection

Synthetic rope slings must be inspected by a competent person each day before being used and regularly during use. Slings must be removed from service when any of the following conditions exist:

- Cuts, gouges, wear, etc. reducing the effective diameter of the rope by more than 10%.
- Powdered fiber between strands.
- Broken, cut, melted or charred fibers.
- Chemical or ultraviolet damage.
- Foreign matter that has permeated the rope and attracts and holds grit.
- Kinks or distortion in the rope structure.
- Corrosion, cracks, distortion, or wear of thimbles or other fittings.
- Missing or illegible sling identification.
- Other visible damage that could affect sling strength.

Broken Fibers

Kinks

NYLON ROPE SLINGS • Eye-and-Eye

Rated Capacity in Pounds • Design Factor = 5

Nominal Rope Diameter (inches)	Vertical	Choker	BASKET HITCH			
			Vertical Basket	60 degree	45 degree	30 degree
1/2	1100	830	2200	1900	1600	1100
9/16	1400	1100	2800	2400	2000	1400
5/8	1800	1400	3600	3100	2500	1800
3/4	2600	2000	5200	4500	3700	2600
7/8	3500	2600	7000	6100	4900	3500
1	4400	3300	8800	7600	6200	4400
1 1/8	5700	4300	11400	9900	8100	5700
1 1/4	7000	5300	14000	12100	9900	7000
1 5/16	7700	5800	15400	13300	10900	7700
1 1/2	9700	7300	19400	16800	13700	9700
1 5/8	11500	8600	23000	19900	16300	11500
1 3/4	13200	9900	26400	22900	18700	13200
2	16900	12700	33800	29300	23900	16900
2 1/8	19100	14300	38200	33100	27000	19100
2 1/4	21400	16100	42800	37100	30300	21400
2 1/2	26300	19700	52600	45600	37200	26300
2 5/8	28800	21600	57600	49900	40700	28800
3	37100	27800	74200	64300	52500	37100

- The use of slings with angles less than 30 degrees from the horizontal (or more than 60 degrees from the vertical) is not recommended.
- The contact surface where the eye is placed must have a diameter of curvature at least double the diameter of the sling.
- Contact surface where sling body is placed must have a diameter of curvature at least 8 times the diameter of the sling.

NYLON ROPE SLINGS • Endless

Rated Capacity in Pounds • Design Factor = 5

Nominal Rope Diameter (inches)	Vertical	Choker	BASKET HITCH			
			Vertical Basket	60 degree	45 degree	30 degree
1/2	2000	1500	4000	3500	2800	2000
9/16	2600	2000	5200	4500	3700	2600
5/8	3200	2400	6400	5500	4500	3200
3/4	4600	3500	9200	8000	6500	4600
7/8	6200	4700	12400	10700	8800	6200
1	7900	5900	15800	13700	11200	7900
1 1/8	10100	7600	20200	17500	14300	10100
1 1/4	12400	9300	24800	21500	17500	12400
1 5/16	13700	10300	27400	23700	19400	13700
1 1/2	17400	13100	34800	30100	24600	17400
1 5/8	20500	15400	41000	35500	29000	20500
1 3/4	23600	17700	47200	40900	33400	23600
2	30200	22700	60400	52300	42700	30200
2 1/8	34100	25600	68200	59100	48200	34100
2 1/4	38300	28700	76600	66300	54200	38300
2 1/2	46900	35200	93800	81200	66300	46900
2 5/8	51400	38600	102800	89000	72700	51400
3	66200	49700	132400	114700	93600	66200

- For an endless sling with a vertical hitch carrying a load of such size as to throw the legs more than 5 degrees off vertical, use rated capacity data for eye-and -eye sling, basket hitch, and corresponding leg angles.
- The use of slings with angles less than 30 degrees from the horizontal (or more than 60 degrees from the vertical) is not recommended.
- Contact surface where sling body is placed must have a diameter of curvature at least 8 times the diameter of the sling.

POLYESTER ROPE SLINGS • Eye-and-Eye

Rated Capacity in Pounds • Design Factor = 5

Nominal Rope Diameter (inches)	Vertical	Choker	BASKET HITCH			
			Vertical Basket	60 degree	45 degree	30 degree
1/2	1000	750	2000	1700	1400	1000
9/16	1300	980	2600	2300	1800	1300
5/8	1600	1200	3200	2800	2300	1600
3/4	2200	1700	4400	3800	3100	2200
7/8	3000	2300	6000	5200	4200	3000
1	4000	3000	8000	6900	5700	4000
1 1/8	5000	3800	10000	8700	7100	5000
1 1/4	6000	4500	12000	10400	8500	6000
1 5/16	6500	4900	13000	11300	9200	6500
1 1/2	8400	6300	16800	14500	11900	8400
1 5/8	9900	7400	19800	17100	14000	9900
1 3/4	11400	8600	22800	19700	16100	11400
2	14400	10800	28800	24900	20400	14400
2 1/8	16200	12200	32400	28100	22900	16200
2 1/4	18100	13600	36200	31300	25600	18100
2 1/2	22000	16500	44000	38100	31100	22000
2 5/8	24200	18200	48400	41900	34200	24200
3	31200	23400	62400	54000	44100	31200

- The use of slings with angles less than 30 degrees from the horizontal (or more than 60 degrees from the vertical) is not recommended.
- The contact surface where the eye is placed must have a diameter of curvature at least double the diameter of the sling.
- Contact surface where sling body is placed must have a diameter of curvature at least 8 times the diameter of the sling.

POLYESTER ROPE SLINGS • Endless

Rated Capacity in Pounds • Design Factor = 5

Nominal Rope Diameter (inches)	Vertical	Choker	Vertical Basket	BASKET HITCH 60 degree	45 degree	30 degree
1/2	1800	1400	3600	3100	2500	1800
9/16	2300	1700	4600	4000	3300	2300
5/8	2800	2100	5600	4800	4000	2800
3/4	4000	3000	8000	6900	5700	4000
7/8	5400	4100	10800	9400	7600	5400
1	7100	5300	14200	12300	10000	7100
1 1/8	8900	6700	17800	15400	12600	8900
1 1/4	10600	8000	21200	18400	15000	10600
1 5/16	11600	8700	23200	20100	16400	11600
1 1/2	15100	11300	30200	26200	21400	15100
1 5/8	17600	13200	35200	30500	24900	17600
1 3/4	20400	15300	40800	35300	28800	20400
2	25700	19300	51400	44500	36300	25700
2 1/8	28900	21700	57800	50100	40900	28900
2 1/4	32300	24200	64600	55900	45700	32300
2 1/2	39300	29500	78600	68100	55600	39300
2 5/8	43200	32400	86400	74800	61100	43200
3	55700	41800	111400	96500	78800	55700

- For an endless sling with a vertical hitch carrying a load of such size as to throw the legs more than 5 degrees off vertical, use rated capacity data for eye-and-eye sling, basket hitch, and corresponding leg angles.
- The use of slings with angles less than 30 degrees from the horizontal (or more than 60 degrees from the vertical) is not recommended.
- Contact surface where sling body is placed must have a diameter of curvature at least 8 times the diameter of the sling.

POLYPROPYLENE ROPE SLINGS • Eye-and-Eye

Rated Capacity in Pounds • Design Factor = 5

Nominal Rope Diameter (inches)	Vertical	Choker	BASKET HITCH			
			Vertical Basket	60 degree	45 degree	30 degree
1/2	760	570	1500	1300	1100	750
9/16	920	690	1800	1600	1300	900
5/8	1100	830	2200	1900	1600	1100
3/4	1500	1100	3000	2600	2100	1500
7/8	2100	1600	4200	3600	3000	2100
1	2600	2000	5200	4500	3700	2600
1 1/8	3200	2400	6400	5500	4500	3200
1 1/4	3900	2900	7800	6800	5500	3900
1 5/16	4200	3200	8400	7300	5900	4200
1 1/2	5500	4100	11000	9500	7800	5500
1 5/8	6400	4800	12800	11100	9000	6400
1 3/4	7400	5600	14800	12800	10500	7400
2	9400	7100	18800	16300	13300	9400
2 1/8	10500	7900	21000	18200	14800	10500
2 1/4	11900	8900	23800	20600	16800	11900
2 1/2	14400	10800	28800	24900	20400	14400
2 5/8	16100	12100	32200	27900	22800	16100
3	20500	15400	41000	35500	29000	20500

- The use of slings with angles less than 30 degrees from the horizontal (or more than 60 degrees from the vertical) is not recommended.
- The contact surface where the eye is placed must have a diameter of curvature at least double the diameter of the sling.
- Contact surface where sling body is placed must have a diameter of curvature at least 8 times the diameter of the sling.

POLYPROPYLENE ROPE SLINGS • Endless

Rated Capacity in Pounds • Design Factor = 5

Nominal Rope Diameter (inches)	Vertical	Choker	BASKET HITCH			
			Vertical Basket	60 degree	45 degree	30 degree
1/2	1400	1100	2800	2400	2000	1400
9/16	1600	1200	3200	2800	2300	1600
5/8	2000	1500	4000	3500	2800	2000
3/4	2700	2000	5400	4700	3800	2700
7/8	3700	2800	7400	6400	5200	3700
1	4600	3500	9200	8000	6500	4600
1 1/8	5700	4300	11400	9900	8100	5700
1 1/4	6900	5200	13800	12000	9800	6900
1 5/16	7600	5700	15200	13200	10700	7600
1 1/2	9800	7400	19600	17000	13900	9800
1 5/8	11400	8600	22800	19700	16100	11400
1 3/4	13200	9900	26400	22900	18700	13200
2	16700	12500	33400	28900	23600	16700
2 1/8	18800	14100	37600	32600	26600	18800
2 1/4	21200	15900	42400	36700	30000	21200
2 1/2	25700	19300	51400	44500	36300	25700
2 5/8	28800	21600	57600	49900	40700	28800
3	36600	27500	73200	63400	51800	36600

- For an endless sling with a vertical hitch carrying a load of such size as to throw the legs more than 5 degrees off vertical, use rated capacity data for eye and eye sling, basket hitch, and corresponding leg angles.
- The use of slings with angles less than 30 degrees from the horizontal (or more than 60 degrees from the vertical) is not recommended.

HARDWARE • General Information

Hardware is an integral and important part of a rigging operation. For an operation to be conducted in a safe and ultimately successful manner, personnel must be trained and qualified in the proper use and inspection.

Hardware should be marked with its name, size and rated capacity. Modifications should only be made when approved by the manufacturer, and repairs made in accordance to the manufacturer's instructions.

Consult the manufacturer before using hardware at temperatures above 400°F or below -40°F; for blocks, above 150°F or below 0°F; for swivel hoist rings, above 400°F or below -20°F; for carbon steel eye bolts, above 275°F or below -30°F.

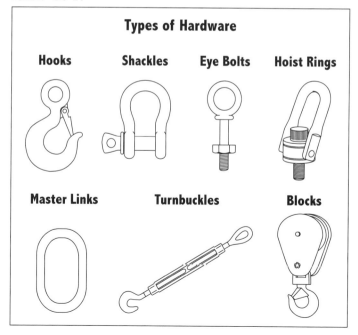

HARDWARE • Hooks

Hooks could very well be the most used type of rigging hardware. They are made in many different sizes and shapes to meet a wide range of applications. They can be attached to load blocks, slings, and other lifting devices such as lifting beams. Hooks must be forged, cast, or die stamped with the manufacturer's ID, rated capacity, and equipped with latches where applicable.

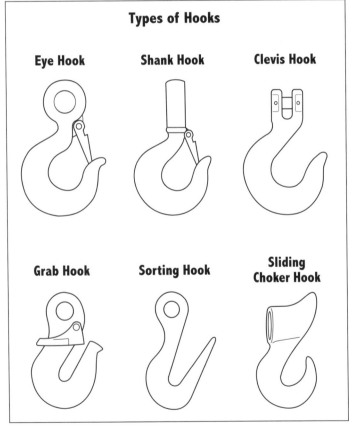

HARDWARE • Hooks

Application

When using two slings placed in a hook, ensure that the included angle between the slings is not greater than 90°. This prevents the slings from coming out of the hook and prevents point loading which reduces hook capacity.

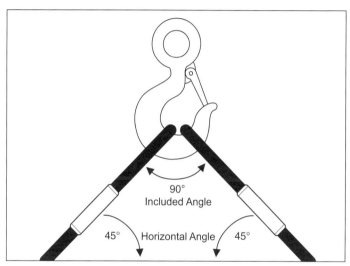

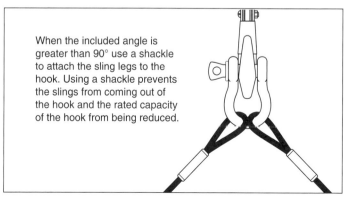

When the included angle is greater than 90° use a shackle to attach the sling legs to the hook. Using a shackle prevents the slings from coming out of the hook and the rated capacity of the hook from being reduced.

HARDWARE • Hooks

Application

Ensure that the hook, not the latch, supports the load. The sling or lifting device must always be seated properly in the bowl of the hook.

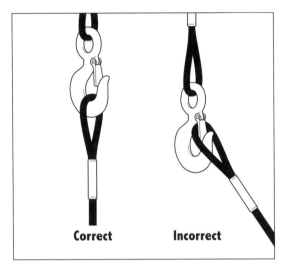

Never side load, back load, or point load a hook. All reduce hook strength and create an unsafe condition. Point loading can reduce hook capacity as much as 60%.

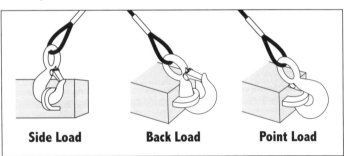

HARDWARE • Hooks

Inspection

Before use, hooks must be inspected by a competent person and removed from service when any of the following conditions exist:

- Manufacturer's identification absent.
- Cracks, nicks or gouges.
- Any twist from plane of unbent hook.
- Latch engagement, damage or malfunction.
- An increase in throat opening exceeding 5% not to exceed 1/4 inch (or as recommended by the manufacturer).
- Wear exceeding 10% of original dimension.
- Damage from heat.
- Unauthorized repairs.

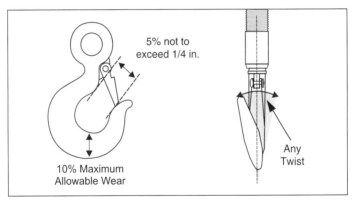

Cracks, nicks, and gouges should be removed by a qualified person (grinding lengthwise, following the contour of the hook) to restore a smooth surface. If removing the damaged area results in a loss of more than 10% of the original dimension, the hook must be replaced. Never repair, alter, or reshape a hook by welding, heating, burning or bending, unless approved by the hook manufacturer.

HARDWARE • Shackles

Shackles, which are normally used to connect two lifting devices, are an essential element of most rigging operations. They should be stamped or embossed with their rated capacity and size. The size of a shackle is determined by the diameter of the body, not by the diameter of the pin.

The two most common types of shackles used in rigging operations are the *anchor type* and *chain type*.

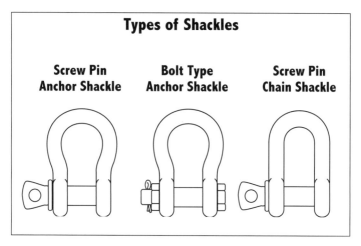

Types of Shackles

Screw Pin Anchor Shackle — Bolt Type Anchor Shackle — Screw Pin Chain Shackle

Specialty shackles are designed to be used for a specific application. One example (at right) is a shackle manufactured predominantly for use with synthetic web slings. This type shackle provides a wider bearing surface, resulting in an increased area for load distribution on the sling, and reduces the tendency of the sling to slide and bunch up.

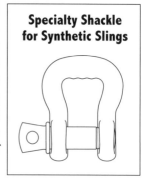

Specialty Shackle for Synthetic Slings

HARDWARE • Shackles

Application

The correct way to use a shackle with a hook is with the shackle pin positioned across the hook. When the bow of the shackle is placed over the hook, the pulling action of the slings can cause the shackle to spread and sling eyes to be damaged.

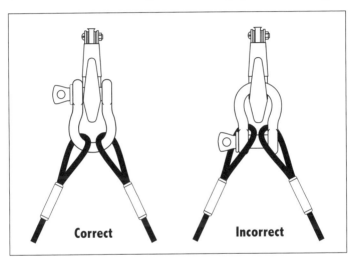

Correct   Incorrect

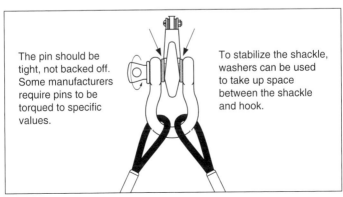

The pin should be tight, not backed off. Some manufacturers require pins to be torqued to specific values.

To stabilize the shackle, washers can be used to take up space between the shackle and hook.

HARDWARE • Shackles

Application

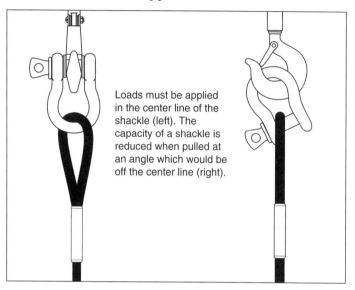

Loads must be applied in the center line of the shackle (left). The capacity of a shackle is reduced when pulled at an angle which would be off the center line (right).

The rated capacity of shackles only applies when they are symmetrically loaded and the included angle between two sling legs is a maximum of 120°. When the angle is greater than 120°, shackle capacity must be reduced.

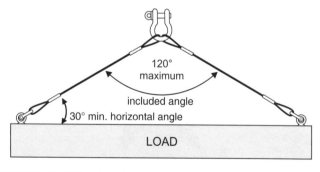

HARDWARE • Shackles

Application

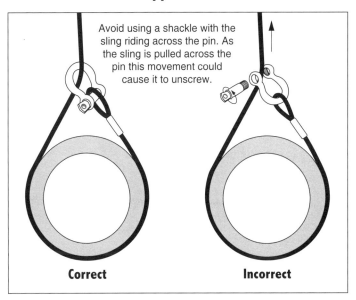

Avoid using a shackle with the sling riding across the pin. As the sling is pulled across the pin this movement could cause it to unscrew.

Correct **Incorrect**

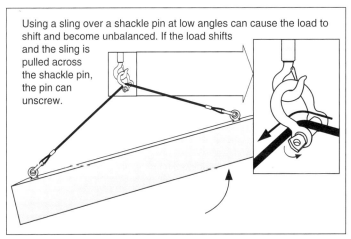

Using a sling over a shackle pin at low angles can cause the load to shift and become unbalanced. If the load shifts and the sling is pulled across the shackle pin, the pin can unscrew.

HARDWARE • Shackles

Application

Never replace a shackle pin with a bolt. The strength of an ordinary bolt is much less than that of a shackle pin.

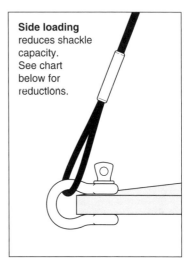

Side loading reduces shackle capacity. See chart below for reductions.

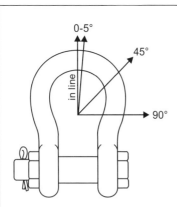

Side Loading Reduction
Screw Pin and Bolt Type Shackles

Angle of Side Load from Vertical In-Line of Shackle	Percent Rated Load Reduction
0 – 5°	0%
6 – 45°	30%
46 – 90°	50%
over 90°	avoid

- In-line load (0°) is applied at right angle (perpendicular) to the pin.
- Do not side load round pin shackles (unthreaded pin without nut).
- Contact manufacturer for side load reduction.

HARDWARE • Shackles

Inspection

Before use, shackles must be inspected visually by a competent person. If any of the following conditions exist, the shackle must be removed from service.

- Manufacturer's name or trademark, capacity and size are illegible.

- Bent or distorted pin and/or body.

- Nicks, gouges or cracks.

- Heat or chemical damage.

- Body spread.

- Shoulder of pin is not flush with shackle body.

- Reduction in diameter of pin and/or body by more than 10%.

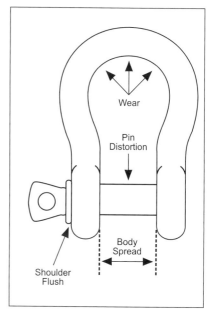

For original dimensions, tolerances, and procedures for making alterations and repairs, consult the shackle manufacturer. Any shackle that has been altered or repaired must be removed from service if manufacturer procedures have not been followed.

HARDWARE • Shackles

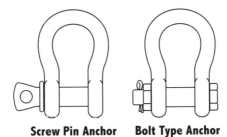

Screw Pin Anchor   **Bolt Type Anchor**

Forged with Alloy Pins

Nominal Shackle Size (inches)	Rated Capacity (lbs)
3/16	660
1/4	1000
5/16	1500
3/8	2000
7/16	3000
1/2	4000
5/8	6500
3/4	9500
7/8	13000
1	17000
1 1/8	19000
1 1/4	24000
1 3/8	27000
1 1/2	34000
1 3/4	50000
2	70000
2 1/4	80000
2 1/2	110000

From the Crosby Group

- Use rated capacity marked on shackle if different from capacities listed above.
- If capacity marking is absent, shackle should be removed from service.

HARDWARE • Eye Bolts

Typically there are two types of eye bolts used in the field: *plain* or *non-shoulder*, and *shoulder*. It is recommended that only forged eye bolts be used for lifting, and it's important to remember that only shoulder eye bolts be used for angular loading. The strength of non-shoulder eye bolts is greatly diminished when loaded at angles. To avoid this, only *shoulder* type eye bolts should be purchased.

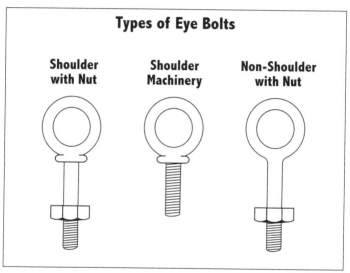

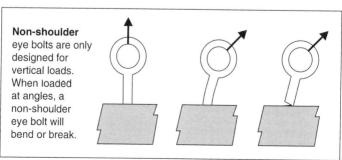

Non-shoulder eye bolts are only designed for vertical loads. When loaded at angles, a non-shoulder eye bolt will bend or break.

HARDWARE • Eye Bolts

Installation

When installing eye bolts the manufacturer's instructions and procedures must be followed. Hex nuts must be tightened securely (or torqued if required) with shoulder flush against the load. Tapped holes, when used, must be drilled and threaded to minimum depth.

The eye bolt must also be positioned so loading can be taken in the plane of the eye. To accomplish this, washers can be used to take up excess space between the load and nut. When using machinery type eye bolts, shims can be used.

The load must also be strong enough to support the forces created by its own weight and angular loading. When using eye bolts with materials other than steel, consult the eye bolt manufacturer.

Shoulder eye bolts must always be positioned to take the load in the plane of the eye. An eye bolt that is "turned to the side" will have less capacity and may experience damage and failure when a load is lifted.

Right **Wrong**

HARDWARE • Eye Bolts

Installation

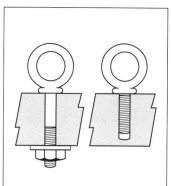

Eye bolts must be tightened securely and torqued if required by the manufacturer, with the shoulder flush against the load.

For angular lifts, the shoulder must be flush, making full contact with the load. Otherwise, only vertical lifts are allowed.

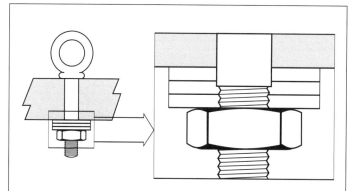

If the nut cannot be tightened securely against the load, washers can be used to take up excess space between the load and nut.

To ensure that the nut tightens securely, spacers must extend beyond the threaded portion of the eye bolt.

HARDWARE • Eye Bolts

Installation

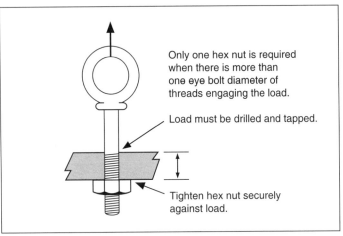

Only one hex nut is required when there is more than one eye bolt diameter of threads engaging the load.

Load must be drilled and tapped.

Tighten hex nut securely against load.

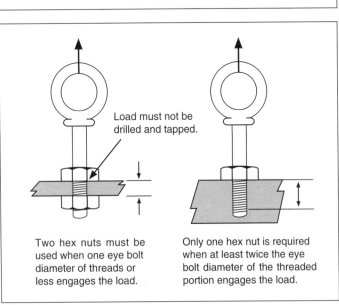

Load must not be drilled and tapped.

Two hex nuts must be used when one eye bolt diameter of threads or less engages the load.

Only one hex nut is required when at least twice the eye bolt diameter of the threaded portion engages the load.

HARDWARE • Eye Bolts

Installation

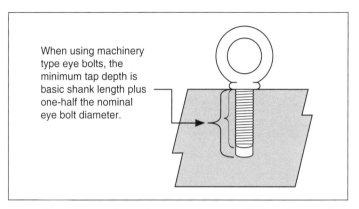

When using machinery type eye bolts, the minimum tap depth is basic shank length plus one-half the nominal eye bolt diameter.

When the receiving hole has been drilled and tapped to correct size and depth, the plane of the eye must be aligned (parallel) with the sling line. If necessary, shims or washers can be used to enable the eye to be brought into the proper alignment. The table below can be used to select the correct shim thickness.

Eye Bolt Size (inches)	Shim Thickness Required to Change Rotation 90°
1/4	.0125
5/16	.0139
3/8	.0156
1/2	.0192
5/8	.0227
3/4	.0250
7/8	.0270
1	.0312
1 1/4	.0357
1 1/2	.0427

HARDWARE • Eye Bolts

Application

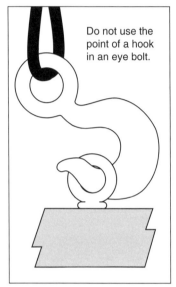

Do not use the point of a hook in an eye bolt.

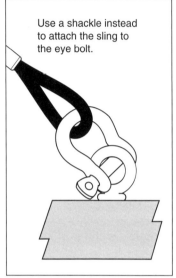

Use a shackle instead to attach the sling to the eye bolt.

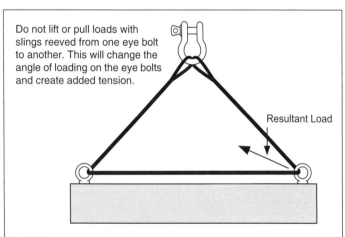

Do not lift or pull loads with slings reeved from one eye bolt to another. This will change the angle of loading on the eye bolts and create added tension.

Resultant Load

HARDWARE • Eye Bolts

Application

When eye bolts are loaded at angles, the increased tension and bending action causes their capacity to be greatly reduced. It is particularly important when eye bolts are loaded at 60° and below to ensure that the manufacturer's instructions are followed and the rated capacities provided are not exceeded.

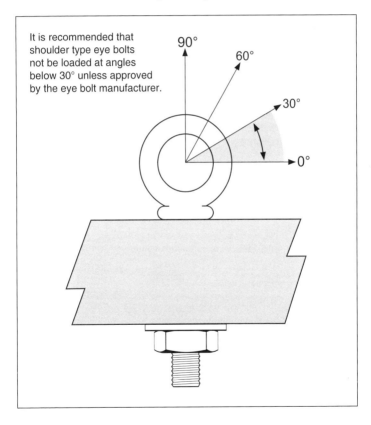

HARDWARE • Eye Bolts

Inspection

Before use, eye bolts must be inspected visually by a competent person. If any of the following conditions exist, the eye bolt must be removed from service.

- Bent or distorted eye or shank.
- Nicks and gouges.
- Obvious wear.
- Worn, corroded and/or distorted threads.
- Heat damage.
- Manufacturer's name or trademark, size or capacity, and grade illegible.

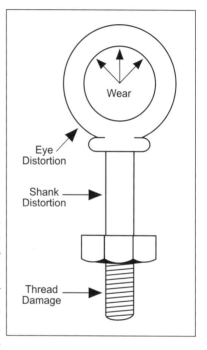

In addition, tapped receiving holes must be cleaned and inspected for thread wear and deterioration. Any alteration or repair to eye bolts, such as grinding, machining, welding, notching, stamping, etc. is not permissible. Eye bolts which have visible signs that alterations or repairs have been made must be removed from service and should be destroyed.

HARDWARE • Eye Bolts

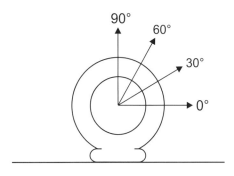

Forged Eye Bolts • Shoulder Type
Rated Capacities in Pounds

Nominal Size (inches)	90 degree	60 degree	30 degree	0 degree
1/4	400	75	NR	NR
5/16	680	210	NR	NR
3/8	1000	400	220	180
1/2	1840	850	520	440
5/8	2940	1410	890	740
3/4	4340	2230	1310	1140
7/8	6000	2960	1910	1630
1	7880	3850	2630	2320
1 1/4	12600	6200	4125	3690
1 1/2	18260	9010	6040	5460

From the Forged Eyebolt Institute

HARDWARE • Hoist Rings

Installation

The following instructions must be followed when installing hoist rings, including any instructions provided by the manufacturer.

- Retention nuts, when used, must have full thread engagement. For the rated capacity to apply, SAE Grade 8 standard hex or equivalent must be used.
- Spacers must not be used between the bushing flange and the mounting surface.
- Contact surface must be flush and in full contact with the hoist ring bushing mating surface.
- Drilled and tapped hole must be 90° to the load surface.
- Using a torque wrench, install hoist ring to recommended torque provided by the manufacturer.

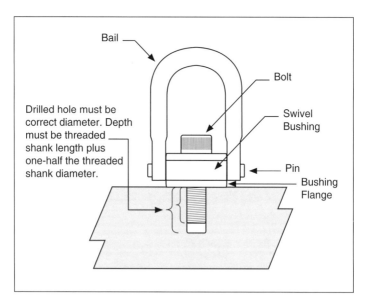

HARDWARE • Hoist Rings

Application

Unlike eye bolts, the rated capacity of hoist rings is not reduced when loaded at angles.

For example, a hoist ring with a rated capacity of 2,500 lbs maintains this capacity even when loaded at an angle. However, an eye bolt with a vertical rated capacity of 6,000 lbs decreases to 2,960 lbs when loaded at a 60° angle, and decreases further to 1,910 lbs when loaded at a 30° angle.

It is important to remember that when hoist rings are loaded at angles (see illustration below) additional tension is created. This tension plus the actual load weight must not exceed the rated capacity of the hoist rings.

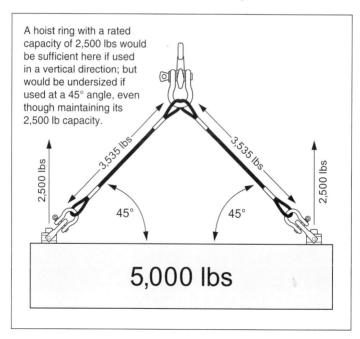

HARDWARE • Hoist Rings

Inspection

Before use, hoist rings must be inspected by a competent person. If any of the following conditions exist, the hoist ring must be removed from service.

- Corrosion, wear or damage.
- Bail does not move freely (it should pivot 180° and swivel 360°).
- Bail is bent, twisted, or elongated.
- Threads on the shank and receiving holes are unclean, damaged, or do not fit properly.
- Heat damage.
- Manufacturer's name or trademark, capacity and torque value are illegible.

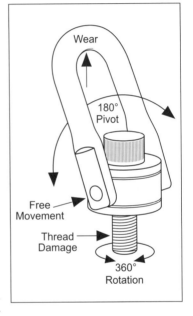

Tapped receiving holes must be cleaned and inspected for thread wear and deterioration.

Any alteration or repair to hoist rings, such as grinding, machining, welding, notching, stamping, etc. is not permissible. Hoist rings which have visible signs that alterations or repairs have been made must be removed from service and should be destroyed.

HARDWARE • Hoist Rings

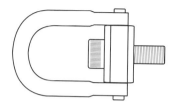

UNC Threads • Working Load Limit in Pounds

Bolt Diameter (inches)	Bolt Length (inches)	Ring Diameter (inches)	Torque (ft lbs)	Working Load Limit (pounds)
5/16	1.50	0.38	7	800
3/8	1.50	0.38	12	1000
1/2	2.00	0.75	28	2500
1/2	2.50	0.75	28	2500
5/8	2.00	0.75	60	4000
5/8	2.75	0.75	60	4000
3/4	2.25	0.75	100	5000
3/4	2.75	0.75	100	5000
3/4	2.75	1.00	100	7000
3/4	3.50	1.00	100	7000
7/8	2.75	1.00	160	8000
7/8	3.50	1.00	160	8000
1	4.00	1.00	230	10000
1 1/4	4.50	1.25	470	15000
1 1/2	6.50	1.75	800	24000
2	6.50	1.75	1100	30000

From the Crosby Group

- Tightening torque values shown are based upon threads being clean, dry and free of lubrication.

- Long bolts are designed to be used with soft metal (i.e., aluminum) work piece. While the long bolts may also be used with ferrous metal (i.e., steel and iron) work piece, short bolts are designed for ferrous work pieces only.

HARDWARE • Master Links

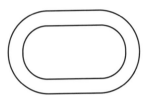

Alloy Steel

Size (inches)	Working Load Limit (pounds)	Weight (pounds)
1/2	4920	0.82
5/8	6600	1.52
3/4	10320	2.07
1	24360	4.85
1 1/4	35160	9.57
1 1/2	47880	16.22
1 3/4	62520	25.22
2	97680	37.04
2 1/4	119400	54.10
2 1/2	147300	67.75
2 3/4	178200	87.80
3	228000	115
3 1/4	262200	145
3 1/2	279000	200
3 3/4	336000	198
4	373000	228
4 1/4	354000	302
4 1/2	360000	342
4 3/4	389000	436

From the Crosby Group

- Working Load Limit is based on single leg sling. Minimum Ultimate Load is 5 times Working Load Limit.
- Sizes 2 inches to 4 3/4 inches are welded.

HARDWARE • Turnbuckles

Turnbuckles come with three types of end fittings: *eye, hook,* and *jaw.* Both ends of a turnbuckle can have the same type end fitting or any two of the three. To prevent rotation, turnbuckles may also be equipped with locknuts. When used in hoisting and rigging applications, turnbuckles should be made from alloy steel or the equivalent, and not welded. They must also be used in a straight or in-line pull only.

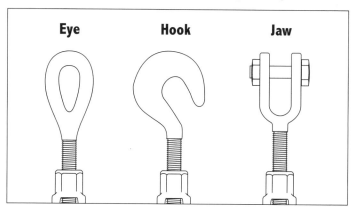

Eye   **Hook**   **Jaw**

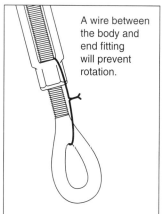

A wire between the body and end fitting will prevent rotation.

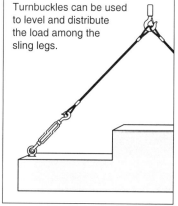

Turnbuckles can be used to level and distribute the load among the sling legs.

HARDWARE • Turnbuckles

Inspection

Before use, turnbuckles must be inspected by a competent person and removed from service when any of the following conditions exist. Turnbuckles must not be altered or repaired without approval from the manufacturer.

Each turnbuckle must be marked to show:
- manufacturer's name or trademark
- size or capacity
- grade

End fittings: deformation, wear and other damage

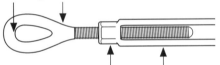

Body: cracks, deformation and other damage

Hooks: throat opening too large; twisting and other damage

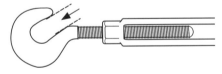

Bolt and Nut: wrong type or size; deformation, thread wear and other damage

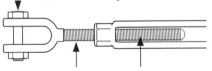

Rod: deformation, thread damage and wear

HARDWARE • Turnbuckles

Hook & Hook **Hook & Eye**   **Eye & Eye** **Jaw & Eye** **Jaw & Jaw**

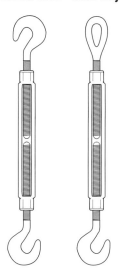
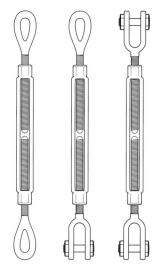

Size (inches)	Rated Capacity (pounds)
1/4	400
5/16	700
3/8	1000
1/2	1500
5/8	2250
3/4	3000
7/8	4000
1	5000
1 1/4	6500
1 1/2	7500

From the Columbus McKinnon Corp.

Size (inches)	Rated Capacity (pounds)
1/4	500
5/16	800
3/8	1200
1/2	2200
5/8	3500
3/4	5200
7/8	7200
1	10000
1 1/4	15200
1 1/2	21400
1 3/4	28000
2	37000
2 1/2	60000
2 3/4	75000

From the Crosby Group

HARDWARE • Blocks

A variety of blocks are available for lifting and moving loads, all specifically designed for use in different applications. *Wire rope blocks* and *crane blocks* are well suited for lifting heavy loads and work well in high speed applications. *Snatch blocks,* which are used to change direction of the wire rope, are intended for intermediate use and slower line speeds. Their side plates open up, allowing the wire rope to be easily attached instead of having to be threaded. *Tackle blocks,* as opposed to other type blocks, are designed to lift light loads using fiber or synthetic ropes.

The block manufacturer should be consulted when rigging blocks are to be used in temperatures over 150°F or below 0°F.

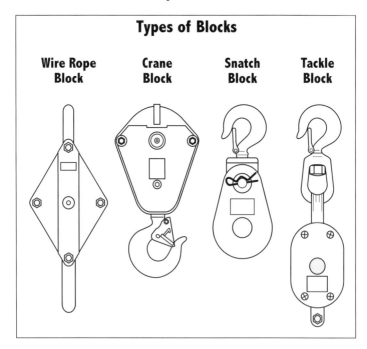

Types of Blocks: Wire Rope Block, Crane Block, Snatch Block, Tackle Block

HARDWARE • Blocks

Reeving and Mechanical Advantage

Mechanical advantage is the leverage gained by reeving a traveling block with multiple parts of hoist line. The more parts used, the more leverage gained and the less line pull required to lift or move the load.

There is no mechanical advantage when using a single part load line; therefore, line pull is equal to the load weight plus any force required to overcome block friction.

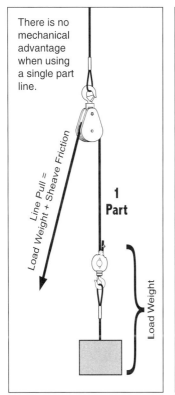

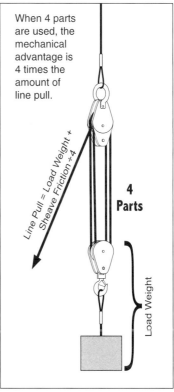

HARDWARE • Blocks

Single Part Load Line System

When single sheave blocks are used to change direction of the load line, the total load on the blocks, rigging, and supporting structure can be much greater than the actual load being lifted. The total load will vary depending on (1) the weight of the load being lifted, (2) the angle between the incoming and departing load line at the block, and (3) the amount of force (line pull) required to overcome sheave friction.

The total load is determined by multiplying the proper *angle factor* (see next page) by the load weight plus the force required to overcome sheave friction.

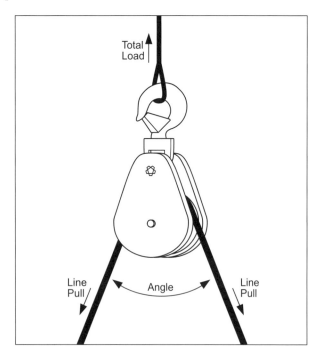

HARDWARE • Blocks

Angle Factors

To determine the total load on the block, rigging, and supporting structure, multiply the angle factor by the line pull.

The amount of line pull required to lift the load will be equal to the load weight plus any force required to overcome sheave friction. For maintained sheaves, friction loss is about 5% per sheave with bushings and about 3% for sheaves with bearings.

Angle Factor Multipliers

Angle	Factor
0°	2.00
10°	1.99
20°	1.97
30°	1.93
40°	1.87
45°	1.84
50°	1.81
60°	1.73
70°	1.64
80°	1.53
90°	1.41
100°	1.29
110°	1.15
120°	1.00
130°	0.84
135°	0.76
140°	0.68
150°	0.52
160°	0.35
170°	0.17
180°	0.00

Total Load = 1,410 lbs + Sheave Friction

Line Pull — 90° (Factor = 1.41) — 1,000 lb Load

Total Load = 760 lbs + Sheave Friction

Line Pull — 135° (Factor = 0.76) — 1,000 lb Load

HARDWARE • Blocks

Hoisting Loads

The example below illustrates a hoisting system with a block reeved with two parts of load line, giving the hoisting system a mechanical advantage of 2. Excluding sheave friction, a line pull of 500 lbs is required to lift the 1,000 lb load.

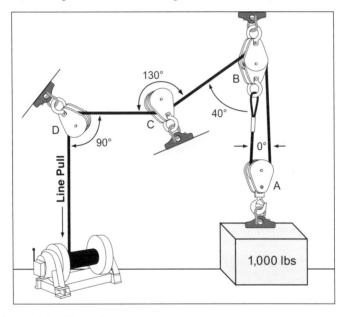

Determining line pull: 1,000 lb load ÷ 2 parts of line = 500 lbs line pull

Calculating Load on Blocks:

Load on Block A = 500 lbs x 2 parts = 1,000 lbs

Load on Block B = 500 lbs (dead end load)
                        + 500 lbs x 1.87 (40° angle factor) = 1,435 lbs

Load on Block C = 500 lbs x 0.84 (130° angle factor) = 420 lbs

Load on Block D = 500 lbs x 1.41 (90° angle factor) = 705 lbs

HARDWARE • Blocks

Hoisting Loads

The example below illustrates a hoisting system with blocks A and B reeved with four parts of load line, giving the hoisting system a mechanical advantage of 4. Excluding sheave friction, a line pull of 875 lbs is required to lift the 3,500 lb load.

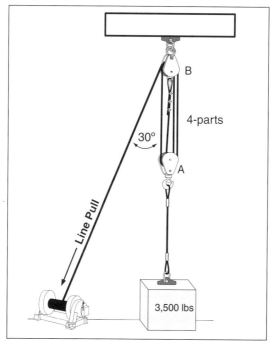

Determining line pull: 3,500 lb load ÷ 4 parts of line = 875 lb line pull

Calculating Load on Blocks:

Load on Block A = 875 lbs x 4 parts = 3,500 lbs

Load on Block B = 875 lbs x 3 parts = 2,625 lbs
                    + 875 lbs x 1.93 (30° angle factor) = 4,314 lbs

HARDWARE • Blocks

Hoisting Loads

The example below illustrates a hoisting system with blocks A and B reeved with three parts of load line and block C reeved with one part of load line, giving the hoisting system a mechanical advantage of 4. Excluding sheave friction, a line pull of 2,500 lbs is required to lift the 10,000 lb load.

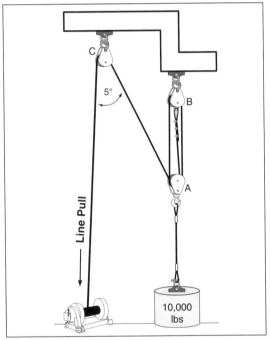

Determining line pull: 10,000 lb load ÷ 4 parts of line = 2,500 lb line pull

Calculating Load on Blocks:
Load on Block A = 2,500 lbs x 4 parts = 10,000 lbs

Load on Block B = 2,500 lbs x 3 parts = 7,500 lbs

Load on Block C = 2,500 lbs x 2.00 (5° angle factor) = 5,000 lbs

HARDWARE • Blocks

Moving Loads Horizontally

To pull or push a load across a surface, a certain resistance must be overcome. The amount of resistance is referred to as the *coefficient of friction* (CF) which can vary depending on the type of load and surface. To determine the force or pull required (P) to move the load, the load weight (L) is multiplied by the *coefficient of friction* (CF). Therefore the formula would be: P = L x CF.

Surfaces	Coefficient of Friction
Concrete on Concrete	0.65
Metal on Concrete	0.60
Wood on Wood	0.50
Wood on Concrete	0.45
Wood on Metal	0.30
Cast Iron on Steel	0.25
Lubricated Surface	0.15
Steel on Steel	0.10
Load on Rollers	0.05
Load on Ice	0.01
Load on Air	0.002

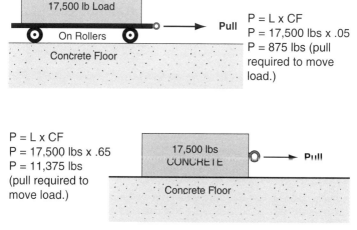

P = L x CF
P = 17,500 lbs x .05
P = 875 lbs (pull required to move load.)

P = L x CF
P = 17,500 lbs x .65
P = 11,375 lbs (pull required to move load.)

HARDWARE • Blocks

Moving Loads Horizontally

The example below illustrates a block of concrete required to be pulled across a concrete floor. Considering the coefficient of friction of concrete, a line pull of 2,600 lbs is required. (Sheave friction not included).

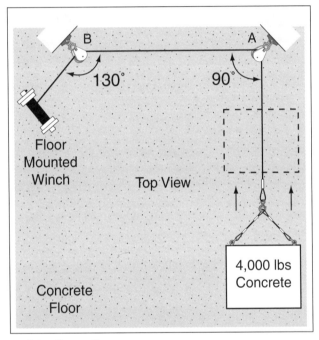

Determining line pull:
4,000 lb load ÷ 1 part x 0.65 (Friction Factor) = 2,600 lbs line pull.

Calculating Load on Blocks:

Load on Block = Line Pull x Angle Factor
Load on Block A = 2,600 lbs x 1.41 (90° angle factor) = 3,666 lbs
Load on Block B = 2,600 lbs x 0.84 (130° angle factor) = 2,184 lbs

HARDWARE • Blocks

Moving Loads Horizontally

The example below illustrates a press on rollers required to be pulled across a concrete floor. Considering the coefficient of friction of rollers on concrete, a line pull of 1,875 lbs is required. (Sheave friction not included).

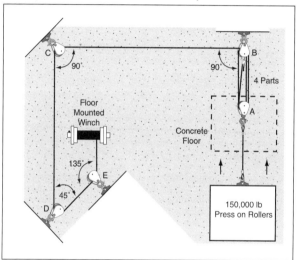

Determining line pull: Load Weight ÷ Parts of Line x Friction Factor

150,000 lb load ÷ 4 parts x 0.05 (Friction Factor) = 1,875 lb line pull

Calculating Load on Blocks:

Load on Block A = 1,875 lbs x 4 parts = 7,500 lbs

Load on Block B = 1,875 lbs x 3 parts = 5,625 lbs
                  1,875 lbs x 1.41 (90° angle factor) = 2,644 lbs
                  5,625 lbs + 2, 644 lbs = 8,269 lbs

Load on Block C = 1,875 lbs x 1.41 (90° angle factor) = 2,644 lbs

Load on Block D = 1,875 lbs x 1.84 (45° angle factor) = 3,450 lbs

Load on Block E = 1,875 lbs x .76 (135° angle factor) = 1,425 lbs

HARDWARE • Blocks

Moving Loads Horizontally

The formula for calculating the length of ramp (incline distance) of a right triangle is $L = \sqrt{(H^2 + R^2)}$. This formula can be used when determining the force or pull required to move a load horizontally up an incline ramp.

Calculations based on:
Length - 2 places   Weight - Next pound

$L = ?$
$H = 4$ ft.
$R = 10$ ft.

$L = \sqrt{(H^2 + R^2)}$
$L = \sqrt{(4^2 + 10^2)}$
$L = \sqrt{(16' + 100')} = 116'$
$L = \sqrt{116'} = 10.77$ ft.

Determining Pull:

Legend (Terms)	Legend (Terms)
R = Run (Horizontal Distance)	W = Weight of Load
H = Height	CF = Coefficient of Friction
L = Length of Ramp	P = Pull required to move load

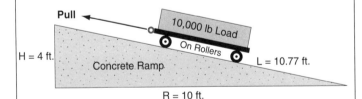

Pull ← 10,000 lb Load On Rollers
H = 4 ft.   Concrete Ramp   L = 10.77 ft.
R = 10 ft.

Formula

$[CF \times W \times (R/L)] + [(H/L) \times W] = P$
.05 friction x 10,000 lbs x (10'/10.77') = 465 lbs
(H/L) x W = (4'/10.77') x 10,000 lbs = 3,715 lbs
465 lbs + 3,715 lbs = 4,180 lb pull required

HARDWARE • Blocks

Inspection

All blocks should be inspected by a competent person before being used, and periodically thereafter. The items illustrated below should be examined during the inspection, and any unsafe condition corrected before the block is put back into service. Under certain conditions, such as use in a corrosive environment or a manufacturer's requirement, blocks may have to be disassembled for a more detailed inspection.

Blocks must not be altered or modified unless approved by the manufacturer. Any repairs must be made in accordance with the manufacturer's instructions.

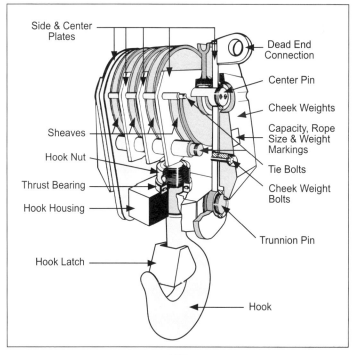

HARDWARE • Lever

Leverage

The example below illustrates one side of a load to be lifted by a wheeled steel lever. The formula for calculating the effort or handle force required to lift one side of the load is: $F_1 = (F_2 \times D_2) \div D_1$.

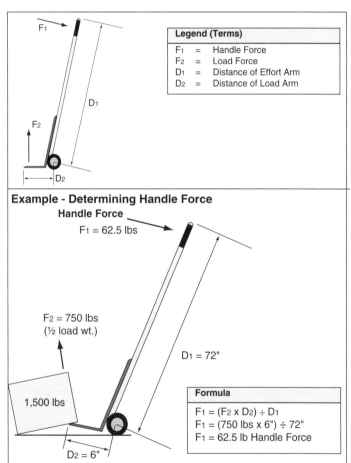

Legend (Terms)		
F_1	=	Handle Force
F_2	=	Load Force
D_1	=	Distance of Effort Arm
D_2	=	Distance of Load Arm

Example - Determining Handle Force

Handle Force
$F_1 = 62.5$ lbs

$F_2 = 750$ lbs (½ load wt.)

1,500 lbs

$D_1 = 72"$

$D_2 = 6"$

Formula
$F_1 = (F_2 \times D_2) \div D_1$
$F_1 = (750 \text{ lbs} \times 6") \div 72"$
$F_1 = 62.5$ lb Handle Force

HARDWARE • Lifting Beams

Two of the most common types of lifting beam used are the *rigid beam* and the *spreader beam.* Rigid beams are composed of a rigid structural member and lift points which can be fixed or adjustable. Spreader beams are composed of a structural member supported by rigging which directs most of the load stress to attachment points. A combination of both types (rigid and spreader beam) is sometimes used.

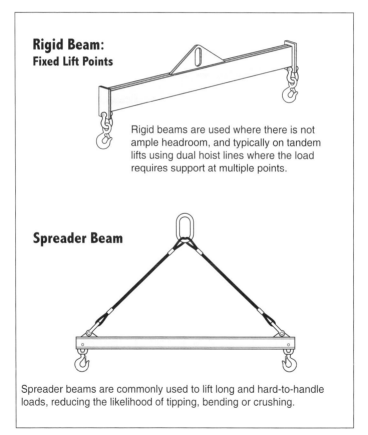

Rigid Beam: Fixed Lift Points

Rigid beams are used where there is not ample headroom, and typically on tandem lifts using dual hoist lines where the load requires support at multiple points.

Spreader Beam

Spreader beams are commonly used to lift long and hard-to-handle loads, reducing the likelihood of tipping, bending or crushing.

HARDWARE • Lifting Beams

Application

Lifting beams enable slings to be used in a vertical configuration, reducing compressive forces on the load. Moreover, they are an efficient way to attach and load slings in highly repetitive operations.

The safe use of a lifting beam requires that the load be distributed and supported by the beam in a manner that enables both the beam and the load to remain level. Side loading is to be avoided, and the beam's rated capacity must never be exceeded. Also, it is important to remember that the weight of the lifting beam is part of the total load weight.

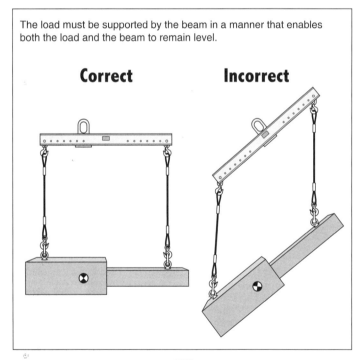

The load must be supported by the beam in a manner that enables both the load and the beam to remain level.

Correct **Incorrect**

HARDWARE • Lifting Beams

Inspection

Lifting beams must be inspected by a competent person before being used, and thereafter at least annually with a record kept. Severity of use and workplace environment may require the beam to be disassembled for a more detailed inspection. If any of the following conditions constitute a hazard, the beam must be removed from service.

- Beam not marked with manufacturer's name and address, serial number, weight, or rated capacity.
- Missing or illegible product safety labels (if applicable).
- Deformation, cracks, or excessive wear of structural members and attachments.
- Loose or missing guards, bolts, fasteners, covers, stops, and nameplates.
- Excessive wear at hoist hooking points and load support hardware and/or pins.

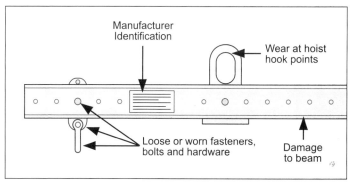

Modifications or alterations must not be made unless first analyzed and approved by the equipment manufacturer or a qualified engineer. Such modifications must also be tested according to applicable ASME standards.

HARDWARE • Lever Operated Hoists

Application

Lever operated hoists, commonly called come-alongs, can be manufactured with roller and welded link chain, as well as wire rope and web strap. They are primarily designed to be used as a pulling device and have limited use in hoisting because of the lever's location.

These devices are often used in rigging applications as an adjustable sling leg to level and position loads at different angles.

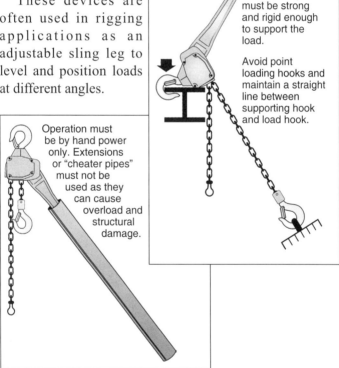

Attachment points must be strong and rigid enough to support the load.

Avoid point loading hooks and maintain a straight line between supporting hook and load hook.

Operation must be by hand power only. Extensions or "cheater pipes" must not be used as they can cause overload and structural damage.

HARDWARE • Lever Operated Hoists

Inspection

Lever operated hoists should be visually inspected before operation and complete inspections, including load limiters for calibration, performed at least annually with a record kept. Covers should be removed for the inspection of internal components.

The following is an example of items to be inspected for unsafe conditions:

- Supporting structure
- Operating mechanisms
- Breaking system
- Hooks & latches
- Hoist lever
- Load chain, wire rope, web strap
- Reeving
- Overtravel restraint
- Warning/caution labels

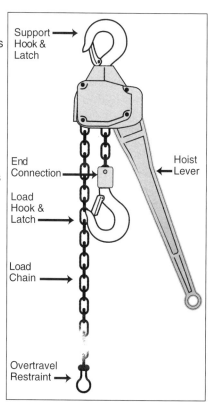

Hoists in which load suspension parts have been altered, replaced or repaired should be load tested with a report placed on file. Test load must not be less than 100% or more than 125% of the rated capacity of the hoist.

HARDWARE • Chain Hoists

Application

Hand chain hoists or chain falls, can be manufactured with welded link or roller chain and can be reeved with single or multiple parts of load chain. They are designed to hang freely and lift loads vertically.

Before lifting loads operators must ensure that there are no twists in the load chain and confirm that the hoist brake will hold the load. This is done by raising and holding the load slightly off the supporting surface before the lift is continued.

Operation must always be by competent personnel and under no circumstances are these hoists to be used where there is not a straight line between the upper and lower hooks.

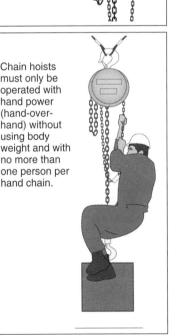

Chain hoists must only be attached to points or supporting structures that have been determined to be capable of supporting the load and any forces created by hoisting.

Chain hoists must only be operated with hand power (hand-over-hand) without using body weight and with no more than one person per hand chain.

HARDWARE • Chain Hoists

Inspection

Hand chain hoists should be visually inspected before operation and complete inspections, including load limiters for calibration, performed at least annually with a record kept. Covers should be removed for the inspection of internal components.

The following is an example of items to be inspected for unsafe conditions:

- Supporting Structure or Trolly
- Operating Mechanisms
- Hoist Braking System
- Load Chain, Reeving, End Connections
- Load Block, Hooks and Latches
- Fasteners, Chain wheels, Shafts, Gears, Pins, Rollers, etc.
- Markings and Warning Labels

Hoists in which load suspension parts have been altered, replaced or repaired should be load tested with a report placed on file. The test load must not be less than 100% or more than 125% of the rated capacity of the hoist.

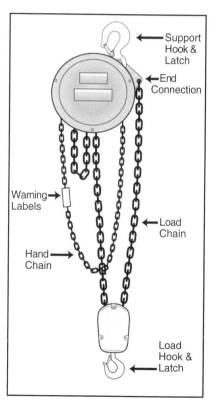

HARDWARE • Load-Indicating Devices

Detachable load-indicating devices (such as crane scales, dynamometers and shackles with load-indicating pins) should be stored in areas where they will not be subject to damage, moisture, corrosive action or extreme temperatures. The manufacturer must be consulted when used in temperatures above 104°F or below 14°F, in chemically active environments, or around electromagnetic or radio frequency interferences.

Inspection

A visual inspection must be performed by a competent person each day before the device is used. A complete inspection must be performed at least annually by a competent person with records kept. Calibration, when required, must be within +/- 2% of the maximum rated load with a written record kept. Load indicating devices must be removed from service when any of the following conditions exist:

- Missing or illegible manufacturer name, trademark, serial number or rated load
- Excessive pitting or corrosion
- Distorted, deformed, elongated, cracked or broken load-bearning components
- Excessive nicks or gouges
- Reduction in orginal dimension inside the load-sensing zone
- 5% reduction in original dimension outside the load-sensing zone
- Illegible display
- Unauthorized welding and heat damage
- Other conditions that could cause doubt as to the continued use

Proof loads must be a minimum of 2 times the rated load.

PROCEDURES • Power Lines

Crane contact with power lines is a major cause of serious injuries and fatalities involving those working with cranes. Overhead power lines must always be considered energized until electrical authorities indicate otherwise. Crane operators and other personnel involved in the operation must not rely on the covering of wires for their protection.

Before any work has begun on a job site containing power lines, the utility company must be contacted to determine if it is feasible for the power lines to be: temporarily diverted around the job site; or de-energized and visibly grounded and appropriately marked at the job site location; or for insulating barriers to be erected to prevent contact between the crane, load, and lines. If none of these are feasible, the following steps must be taken to minimize the hazard of electrocution or serious injury as a result of the crane or load contacting energized power lines.

• An on-site meeting must be held between management of the project and the owner of the power lines (or a designated representative of the electrical utility) to establish procedures to safely complete the operation. Before work begins, these procedures must be communicated to all personnel involved in the operation – including crane operators, signal persons, and riggers, etc.

PROCEDURES • Power Lines

• No part of the crane or load must ever enter the prohibited zone around an energized power line. This zone must be enlarged as electrical potential of the power line increases (see table below). Certain environmental conditions such as fog, smoke, or precipitation may also require this distance to be increased.

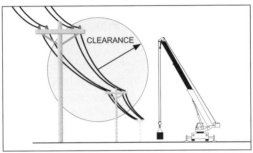

Required clearance for operations near high voltage power lines:	
to 50 kV	10 ft (3.05 m)
over 50 to 200 kV	15 ft (4.60 m)
over 200 to 350 kV	20 ft (6.10 m)
over 350 to 500 kV	25 ft (7.62 m)
over 500 to 750 kV	35 ft (10.67 m)
over 750 to 1000 kV	45 ft (13.72 m)

A 20 foot clearance must be maintained if the voltage, up to 350 kV, is unknown. OSHA 1926.1408(a)(2)
A 50 foot clearance must be maintained if the voltage, over 350 kV, is unknown. OSHA 1926.1409(a)

kV = kilovolt (1000 volts), a unit of electrical potential difference.

• When working around power lines, restrict the working area to essential personnel. A good way to accomplish this is by using barricades.

• Consider erecting guard structures or other highly visible devices around power lines to improve visibility and aid in location of the prohibited zone.

• Consider the use of synthetic slings because they can be less conductive than steel type slings.

• Taglines, when required, must be of a non-conductive type.

PROCEDURES • Power Lines

• Any time a crane is working within a boom's length of the prohibited zone, a qualified signal person must be appointed. The signal person's sole responsibility is to be in constant contact with the operator and to verify that the required clearance is maintained.

• No one may touch any part of the crane or load until the signal person indicates it is safe to do so.

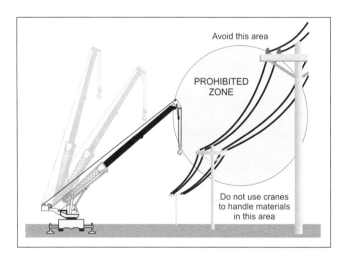

• Materials should not be stored under power lines and cranes should not be used to handle materials under power lines.

• The operation of cranes or handling loads above power lines should be avoided.

• When working close to transmission towers, the crane and rigging can become electrically charged. To reduce the possibility of being shocked, riggers should consider the use of synthetic web slings and insulated gloves when handling suspended loads.

• If insulated links, boom guards, or proximity warning devices are used on cranes, such devices must not be a substitute for the requirements outlined in ASME B30.5. If used, the limitations and testing requirements of these devices must be understood by everyone involved in the operation.

PROCEDURES • Determining Load Weight

Before any rigging operation can begin, the weight of the load to be lifted must be known. Otherwise, it cannot be assumed that the correct rigging equipment has been selected.

The weight of some loads may be easy to calculate because of their simple shape and uniform density. The weight of other loads, because of complex construction, is difficult to determine and may require the assistance of an engineer.

Weights might be obtained from sources such as drawings, shipping documents, and catalogs. Standard tables can also be used for finding the weight of items such as I-beams, bars, pipes, and rods.

When calculating load weight, simplify the process by enlarging the size of irregularly shaped portions of the load into simple shapes such as blocks or cylinders. Doing this will ensure that the estimated weight is higher than the actual weight.

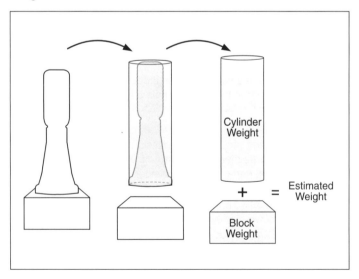

PROCEDURES • Determining Load Weight

Example 1

Shape: Rectangular Block   **Material: Steel**

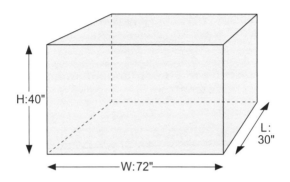

H: Height   W: Width   L: Length

$$\text{Volume} = H \times W \times L$$

= 40" x 72" x 30" = 86,400 cubic inches

Note: 1 cubic foot = 1,728 cubic inches

$$\frac{86,400 \text{ cu in}}{1,728 \text{ cu in / cu ft}} = 50 \text{ cubic feet}$$

$$\text{Load Weight} = \text{Volume} \times \text{Material Weight}$$

Note: Steel weighs 490 pounds per cubic foot

Load Weight = 50 cu ft x 490 lb / cu ft = 24,500 lb

PROCEDURES • Determining Load Weight

Example 2

Shape: Cylinder   **Material: Concrete**

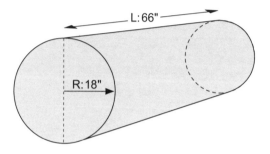

R: Radius   L: Length

$$\text{Volume} = \pi \times R^2 \times L$$

= 3.1416 x (18" x 18") x 66" = 67,180 cubic inches

Note: 1 cubic foot = 1,728 cubic inches

$$\frac{67,180 \text{ cu in}}{1,728 \text{ cu in / cu ft}} = 38.88 \text{ cubic feet}$$

$$\text{Load Weight} = \text{Volume} \times \text{Material Weight}$$

Note: Concrete slag weighs 130 pounds per cubic foot

Weight of Load = 38.88 cu ft x 130 lb / cu ft = 5,055 lb

PROCEDURES • Determining Load Weight
Example 3

Shape: Triangular Prism   Material: Portland Cement (set)

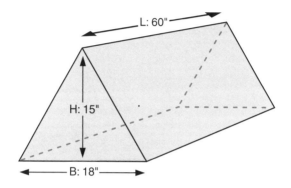

B: Length of base on one end
H: Height from base   L: Length

$$\text{Volume} = \frac{B \times H \times L}{2}$$

$$= \frac{15" \times 18" \times 60"}{2} = 8,100 \text{ cubic inches}$$

Note: 1 cubic foot = 1,728 cubic inches

$$\frac{8,100 \text{ cu in}}{1,728 \text{ cu in / cu ft}} = 4.69 \text{ cubic feet}$$

$$\text{Load Weight} = \text{Volume} \times \text{Material Weight}$$

Note: Portland Cement (set) weighs 183 pounds per cubic foot

Weight of Load = 4.69 cu ft x 183 lb / cu ft = 858.3 lb

PROCEDURES • Determining Load Weight
Weights of Materials and Liquids – lb per cubic ft

Material	Weight	Material	Weight
Aluminum	165	Limestone (solid)	163
Asbestos	153	Lumber: Douglas-fir	34
Asphalt	81	Lumber: Oak	62
Brass	534	Lumber: Pine	45
Brick (Soft)	110	Lumber: Poplar	30
Brick (Common)	125	Lumber: Spruce	28
Brick (Pressed)	140	Lumber: Railroad Ties	50
Bronze	534	Marble	170
Coal	84	Motor Oil	60
Concrete (Slag)	130	Paper	75
Concrete (Reinforced)	150	Petroleum: Crude	55
Copper	556	Petroleum: Gasoline	45
Crushed Rock	95	Portland Cement (Loose)	94
Diesel Fuel	52	Portland Cement (Set)	183
Earth, Dry (Loose)	75	River Sand	120
Earth, Dry (Packed)	95	Rubber	95
Earth, Wet	100	Sand & Gravel (Wet)	125
Glass	161	Sand & Gravel (Dry)	108
Granite	168	Steel	490
Ice	58	Tar	75
Iron	485	Tin	460
Lead	711	Water	65
Lime: Gypsum (Loose)	64	Zinc	440

Weights of Steel and Aluminum Plates
lb per square ft

Plate Size (inches)	Steel	Aluminum
1/8	5	1.75
1/4	10	3.50
1/2	20	7.00
3/4	30	10.50
1	40	14.00

PROCEDURES • Center of Gravity

The *center of gravity* is the point where the entire weight of a body can be considered concentrated so that, if supported at this point, the body would remain in equilibrium in any position. The symbol used for center of gravity (CG) is ⊕.

To avoid having an unbalanced load, the lifting hook must be directly above the center of gravity, which for some loads would require the use of unequal length sling legs.

For loads having a somewhat rectangular shape with the weight of the load concentrated at one end (as in the illustration below right), the center of gravity will be situated toward that end.

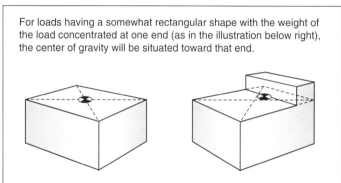

Finding the approximate center of gravity of an irregular shaped load can be done by turning the object into a rectangle and intersecting the lines from opposite corners.

The center of gravity will be close to the place where these diagonal lines intersect.

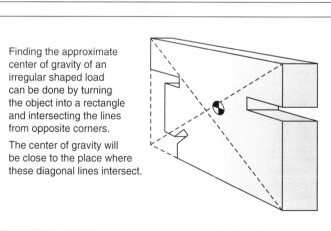

PROCEDURES • Center of Gravity

When a load is suspended, its center of gravity will hang directly below the hook. Using equal length slings on an irregularly shaped load will cause the load to tilt. For the load to hang level, unequal length slings will have to be used.

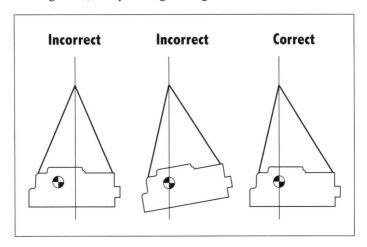

Rigging the load below the center of gravity can result in the load shifting. In order for loads to remain stable, attachment points should be above the center of gravity.

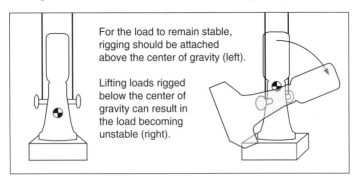

For the load to remain stable, rigging should be attached above the center of gravity (left).

Lifting loads rigged below the center of gravity can result in the load becoming unstable (right).

PROCEDURES • Handling Loads

Softeners

Using slings across sharp edges of a load is a major reason for their failure. To avoid cutting or deforming the sling, always use softeners or padding of sufficient thickness, construction and strength to round out the edge. Steel pipe and wood blocks make excellent softeners.

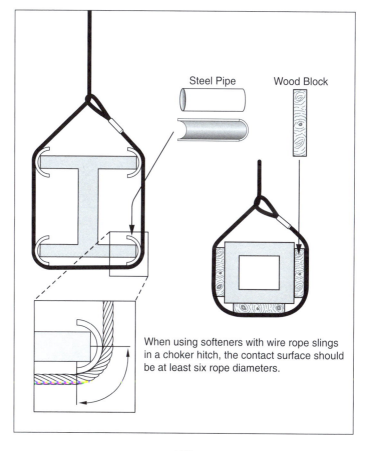

When using softeners with wire rope slings in a choker hitch, the contact surface should be at least six rope diameters.

PROCEDURES • Handling Loads

Attaching Unused Slings

Unused slings, as with all lifting devices and equipment, must not be left hanging free when handling loads. Slings should be attached back to the master link to prevent them from hanging up or striking personnel as the load is being moved.

Hands and clothing must also be kept clear of attachment points, and personnel should never be allowed to ride the load.

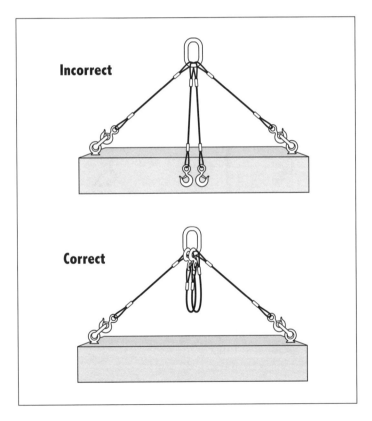

PROCEDURES • Handling Loads

Strength of Loads

Not only do the slings and hardware require sufficient strength, the load itself must be strong enough to withstand the forces from its own weight combined with the compressive forces created by slings used at angles. If strength is insufficient, the load could buckle or be crushed.

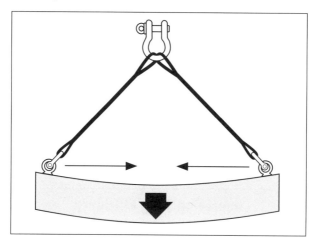

When lifting soft loads (such as crates or wooden boxes) with a basket hitch, spreader beams can be used to prevent damage created from the pressure induced by sling angles. Spreader beams should be slightly longer than the load width and contain no sharp edges.

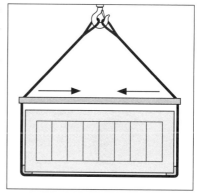

PROCEDURES • Handling Loads

Strength of Loads

Slings should not be reeved through attachments. This would create a resultant load which at least doubles the loading on both the attachments and the load itself.

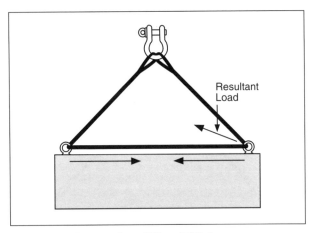

Improving Sling Efficiency

You can improve sling efficiency by placing a wooden block (or blocks) between the hitch and the load, thus increasing the angle between the two choker legs.

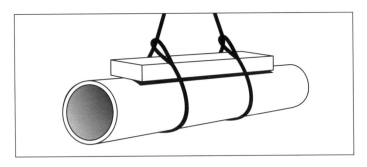

PROCEDURES • Handling Loads

Turning Loads

To turn a load, use a double choker with the sling body passing through the eyes of the sling with the eyes placed in

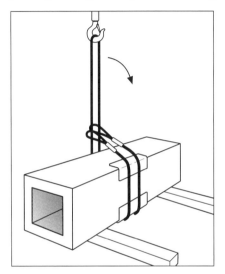

the opposite direction of the turn. To avoid unequal loading of the legs, make sure the center of the sling body is placed over the hook and not the sling eyes.

This method provides good control over the load because its weight is applied against the sling, allowing little or no movement between sling and load.

Turning a load with one hook requires the sling to be attached to the side of the load above the center of gravity. To prevent the load from sliding, the load may have to be simultaneously lifted and moved in the direction of the turn.

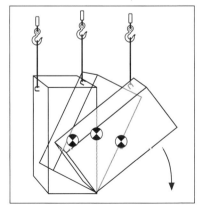

PROCEDURES • Handling Loads

Securing Loads

To prevent loads from becoming dislodged and possibly falling, they must be secured before lifting – especially when lifting loose material and objects such as bricks and blocks.

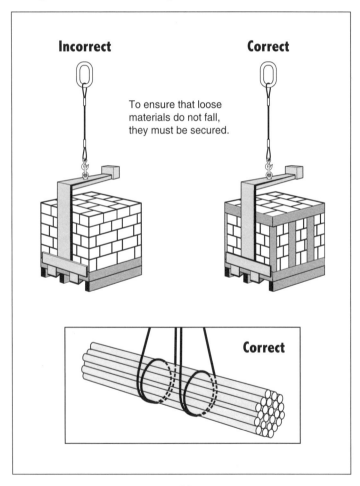

PROCEDURES • Handling Loads

Securing Loads

To avoid having to detach the lifting device from the load, it is common for workers to use "homemade" rigging equipment. One example is to lift wooden trusses with a homemade hook made from a steel rod (or a conventional hook with the latch removed, taped or wired back).

However, there have been cases where a truss has come out of the hook before being secured in place, resulting in a worker being seriously injured or killed. To avoid such accidents, loads must be well secured and properly balanced in the sling or approved lifting device.

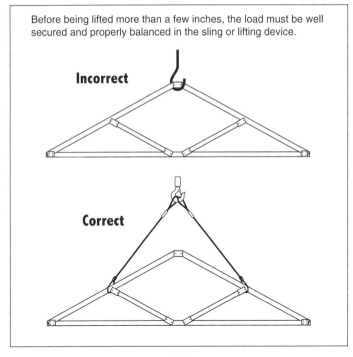

Before being lifted more than a few inches, the load must be well secured and properly balanced in the sling or lifting device.

PROCEDURES • Handling Loads

Tag Lines

A common misconception is that a tag line is required to be used on every load. This can often make controlling the load more difficult and even compromise safety if the tag line becomes tangled with a structure or piece of equipment. However, when a tag line is required to control the load, make sure that it has sufficient strength, no knots and is long enough to keep personnel from under the load.

When working around power lines and other electrical sources, a nonconductive rope should be used.

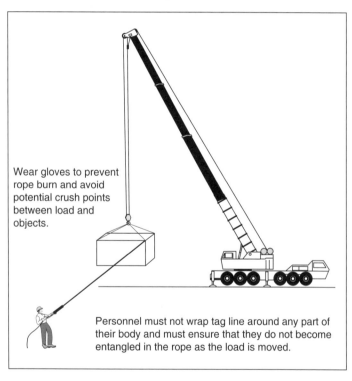

PROCEDURES • Handling Loads

Fall Zone

The fall zone is the area directly beneath the suspended load and the surrounding area in which it is reasonably foreseeable that partially or completely suspended materials could fall in the event of an accident.

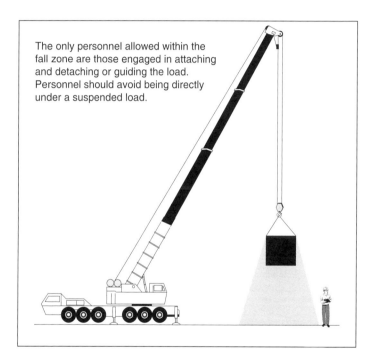

The only personnel allowed within the fall zone are those engaged in attaching and detaching or guiding the load. Personnel should avoid being directly under a suspended load.

PROCEDURES • Handling Loads

Knots

When a rope is formed into a knot, the breaking strength of the rope is reduced approximately 50%, with the knot being the weak point. Therefore, the rope capacity should not be more than 20% of its breaking strength. A good knot is one that can be tied and untied quickly and when tied, will hold. The following are some basic knots that can be used to attach and secure loads in a rigging operation.

Bowline
The *bowline knot* is a loop knot and is used where security is extremely important. Since it will not slip, this makes it one of the most used knots when a loop is required at the end of the rope.

Bowline on a Bight
The *bowline on a bight* is a loop knot used primarily to make a loop in the middle of a rope when the ends are not available.

Bowline on Post
The *bowline on a post* is a bowline knot used when attaching a rope to a post.

PROCEDURES • Handling Loads

Knots

Sheet Bend
A *sheet bend* is a knot used to tie two ropes together whether the same size or different sizes.

Square Knot
A *square knot* is a binding knot and is easily untied. Its primary use is for securing bundles and it can also be used to tie two ropes together.

Clove Hitch
The *clove hitch* is typically used to attach a rope to a post or pipe. It can be tied in the middle or end of the rope.

Double Half Hitch
The *double half hitch* is a half hitch tied twice making it more reliable than the half hitch. It is quickly untied and holds reasonably well when tied properly.

PROCEDURES • Handling Loads

Placement of Loads

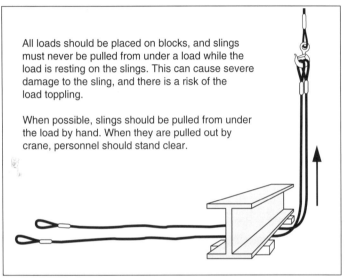

All loads should be placed on blocks, and slings must never be pulled from under a load while the load is resting on the slings. This can cause severe damage to the sling, and there is a risk of the load toppling.

When possible, slings should be pulled from under the load by hand. When they are pulled out by crane, personnel should stand clear.

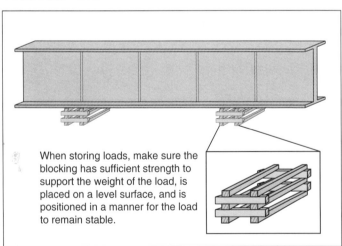

When storing loads, make sure the blocking has sufficient strength to support the weight of the load, is placed on a level surface, and is positioned in a manner for the load to remain stable.

PROCEDURES • Communicating with the Operator

Communicating with the crane operator by hand or voice signal is one of the most important jobs in a crane operation. Since the signal person is in a sense operating the crane, the lift director must only appoint qualified signal persons to direct the operator. To be considered qualified, signal person(s) must be tested and demonstrate that they have a basic understanding of crane operation and limitations; crane dynamics involved in swinging and stopping loads; boom deflection from hoisting loads; and know and have a thorough understanding of standard hand and/or voice signals.

The signal person must be positioned where the operator, path of travel, and location where the load will be placed can clearly be seen. When a lift director is not present, the signal person is responsible for keeping nonessential personnel out of the work area, and must not direct the load over personnel.

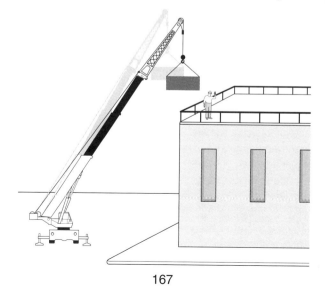

PROCEDURES • Communicating with the Operator

Communication between the signal person and the crane operator must be continuous. If communication is disrupted, crane movements must be stopped until communication is restored and a proper signal is given and understood.

Signals must be discernible or audible and if not understood, no response by the operator should be made. Unless voice communication equipment is used, standard hand signals used to direct the operator must be those prescribed in applicable ASME B30 Standards.

Voice Signals

Before using voice signals, they must be understood and agreed upon between the person directing lifting operations, the crane operator and the signal person. Direction given to the crane operator must be from the operators direction perspective (e.g., swing right) and must contain these elements stated in the following order:

1) function and direction
2) distance and/or speed
3) function stop

Examples of voice signals:
    a) swing right 50 ft, 25 ft, 15 ft, 10 ft, 5 ft, 2 ft, swing stop
    b) load down 100 ft, 50 ft, 40 ft, 30 ft,...2 ft, load stop
    c) load up slow, slow, slow, load stop

Before the operator is signaled to perform more than one crane function at the same time, the lift director must take into consideration the following: complexity of lift, capabilities of crane and ability to communicate the necessary voice signals.

PROCEDURES • Communicating with the Operator

Special Signals

Special signals may be used for operations or crane attachments which are not covered by standard signals. Modifications of the standard voice or hand signals must be agreed upon in advance by the person directing lifting operations, the crane operator, and the signal person. These special signals must not conflict with standard signals.

Audible Emergency Signal

Audible emergency signals can be given by anyone. However, the signals used must be agreed upon for each jobsite location and must not conflict with standard signals. An example would be multiple short audible signals or a continuous audible signal.

Audible Travel Signals for Mobile Cranes

When moving a carrier-mounted crane with two cabs, the following audible travel signals must be used:

- **Stop** - One short audible signal
- **Go ahead** - Two short audible signals
- **Back up** - Three short audible signals

PROCEDURES • Communicating with the Operator

Standard Hand Signals: Mobile Cranes

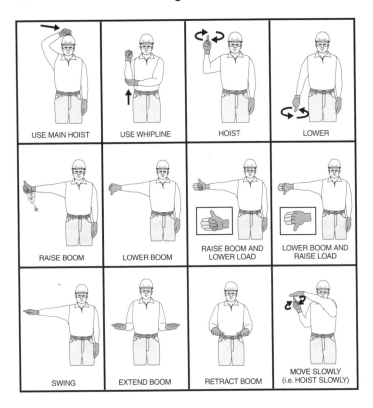

PROCEDURES • Communicating with the Operator

Standard Hand Signals: Mobile Cranes (Continued)

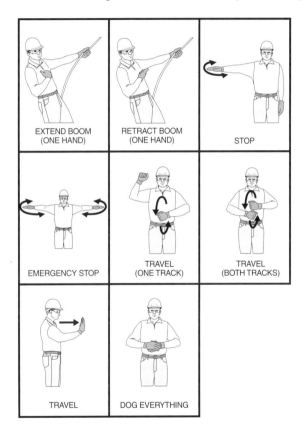

PROCEDURES • Communicating with the Operator

Standard Hand Signals: Tower Cranes

PROCEDURES • Communicating with the Operator

Standard Hand Signals: Overhead and Gantry Cranes

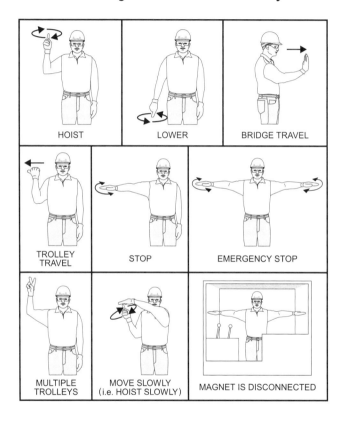

PROCEDURES • Hoisting Personnel

Pre-Lift Considerations

Before any hoisting of personnel by crane begins, the person responsible for the job to be performed must establish that there is no other practical or less hazardous way of doing the job.

That person must then hold a meeting of all personnel involved in the lift. This meeting will include the crane operator, signal person(s) (if considered necessary for the lift), and the employees to be hoisted.

The meeting must be held prior to the trial lift at each new work location, and each time employees are newly assigned to the job.

A specially designed personnel platform, conforming to OSHA 1926.1431 and ASME B30.23 specifications, must be used. Riding the load or the headache ball is *not* permitted. However, OSHA does allow a boatswain's chair for certain work applications — fall protection must be worn.

PROCEDURES • Hoisting Personnel

Platform Specifications

The platform must be designed by a qualified person familiar with structural design and all welding performed by a certified welder. In particular, each platform must have:

- a design factor of 5:1.
- a suspension system to minimize tipping.
- enclosure at least from toeboard to mid-rail.
- points to which fall arrest systems are attached.
- a guardrail and an inside grab rail.
- sufficient headroom for personnel to stand.
- no rough edges which might injure personnel.
- permanent indication of its weight and rated capacity.

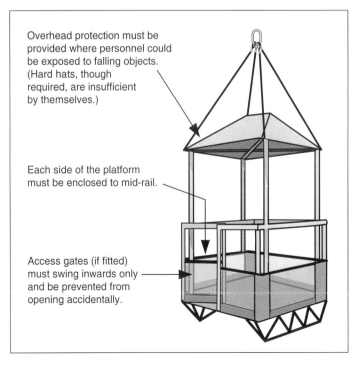

Overhead protection must be provided where personnel could be exposed to falling objects. (Hard hats, though required, are insufficient by themselves.)

Each side of the platform must be enclosed to mid-rail.

Access gates (if fitted) must swing inwards only and be prevented from opening accidentally.

PROCEDURES • Hoisting Personnel

Selection of Rigging

The rigging equipment selected for hoisting personnel must not be used for any other purpose and should be kept apart from other rigging or clearly identified in some way. It must be capable of handling at least 5 times the maximum intended load and 10 times for rotation resistant wire rope slings.

NOTICE
THIS RIGGING TO BE USED FOR PERSONNEL LIFTING **ONLY**. **ALL** OTHER USES PROHIBITED.

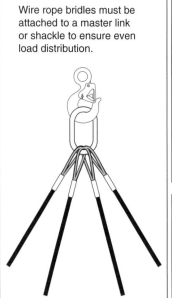

Wire rope bridles must be attached to a master link or shackle to ensure even load distribution.

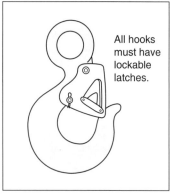

All hooks must have lockable latches.

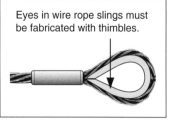

Eyes in wire rope slings must be fabricated with thimbles.

PROCEDURES • Hoisting Personnel

Trial and Test Lifts

The usual correct procedure for setting up the crane must be followed (see companion handbook, *Mobile Cranes*). In particular, the crane must be on a firm surface, level to within 1% and the rated capacity reduced by 50%.

Before any hoisting with a personnel platform begins, a test lift and a trial lift must be conducted.

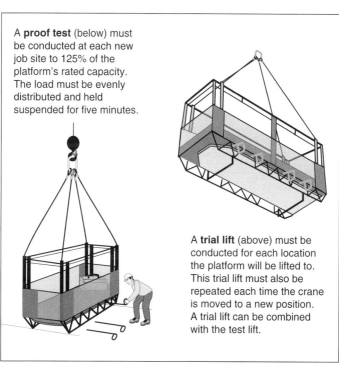

A **proof test** (below) must be conducted at each new job site to 125% of the platform's rated capacity. The load must be evenly distributed and held suspended for five minutes.

A **trial lift** (above) must be conducted for each location the platform will be lifted to. This trial lift must also be repeated each time the crane is moved to a new position. A trial lift can be combined with the test lift.

Note: See OSHA 1926.1431 and ASME B30.23 for more information.

PROCEDURES • Multi-Crane Lifts

Using more than one crane to lift and place a load compounds the risk of the lifting procedure due to the complexity of coordinating the operation. This complexity increases both with the number of cranes lifting the load as well as the maneuvers that must be performed to handle the load. (This is particularly true when utilizing mobile cranes. For example, a simple movement such as booming down with a load – which may be safe and proper for a single crane – can cause a catastrophe if not coordinated properly during a multi-crane lift.)

Consequently, detailed plans must be made, preferably by a qualified engineer, for coordinating every step of a multi-crane lifting procedure. The following issues must be addressed:

- Only one qualified person must direct all operations during the lift to prevent uncoordinated movements.

- All personnel must understand all phases of the operation in addition to their own specific responsibilities.

- Plans must include an accurate determination of the share of the load to be carried by each crane as well as the method by which this load distribution will be controlled during lifting.

- Movements with the load should be planned in simple single steps rather than simultaneously, and made in a slow and controlled manner.

- Hoist lines must remain vertical. (This is absolutely critical with mobile cranes because of the potential for boom collapse from side loading.)

- A reduction of 25% in net capacity for each crane should be considered, although such a reduction alone should never be viewed as adequate preparation for attempting a tandem lift.

- All planning normally required for a single crane lift (such as verifying the adequacy of the supporting surface) becomes more crucial when a multi-crane lift is to be performed.

PROCEDURES • Dual Crane Lifts

Determining Loads on Unequally Loaded Cranes

Example 1:
Both Lift Points at Edge of Load

When both lift points are at the edge of the load, the load on each crane can be determined by
(1) measuring the horizontal distance from the center of gravity of the load to the opposite lift point at the edge of the load,
(2) dividing the result of (1) by the total length of the load between lift points, and
(3) multiplying the result of (2) by the weight of the load.

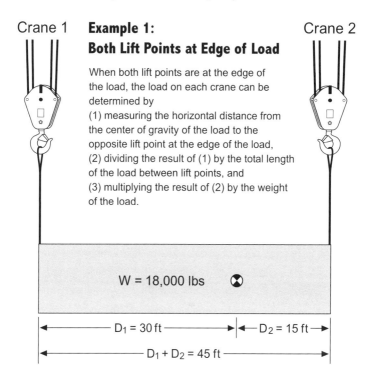

$$\text{Load on Crane 1} = \frac{D_2}{D_1 + D_2} \times W = \frac{15 \text{ ft}}{45 \text{ ft}} \times 18{,}000 \text{ lbs} = 6{,}000 \text{ lbs}$$

$$\text{Load on Crane 2} = \frac{D_1}{D_1 + D_2} \times W = \frac{30 \text{ ft}}{45 \text{ ft}} \times 18{,}000 \text{ lbs} = 12{,}000 \text{ lbs}$$

PROCEDURES • Dual Crane Lifts

Determining Loads on Unequally Loaded Cranes

Example 2: Lift Points Not at Edge of Load

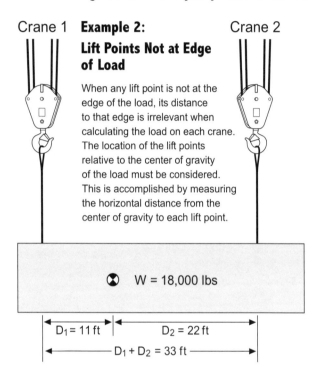

When any lift point is not at the edge of the load, its distance to that edge is irrelevant when calculating the load on each crane. The location of the lift points relative to the center of gravity of the load must be considered. This is accomplished by measuring the horizontal distance from the center of gravity to each lift point.

$$\text{Load on Crane 1} = \frac{D_2}{D_1 + D_2} \times W = \frac{22 \text{ ft}}{33 \text{ ft}} \times 18{,}000 \text{ lbs} = 12{,}000 \text{ lbs}$$

$$\text{Load on Crane 2} = \frac{D_1}{D_1 + D_2} \times W = \frac{11 \text{ ft}}{33 \text{ ft}} \times 18{,}000 \text{ lbs} = 6{,}000 \text{ lbs}$$

PROCEDURES • Dual Crane Lifts

Determining Loads on Unequally Loaded Cranes

Example 3:
Load Not Level; Both Lift Points at Edge of Load

Distances are still measured horizontally when the load is not level. As a result, D_1 and D_2 will be reduced.

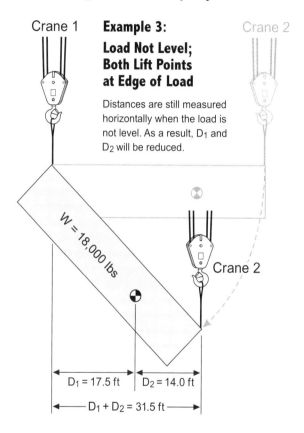

$$\text{Load on Crane 1} = \frac{D_2}{D_1 + D_2} \times W = \frac{14.0 \text{ ft}}{31.5 \text{ ft}} \times 18{,}000 \text{ lbs} = 8{,}000 \text{ lbs}$$

$$\text{Load on Crane 2} = \frac{D_1}{D_1 + D_2} \times W = \frac{17.5 \text{ ft}}{31.5 \text{ ft}} \times 18{,}000 \text{ lbs} = 10{,}000 \text{ lbs}$$

CRANE INSTITUTE OF AMERICA

Training and Certification/Qualification

Crane Institute of America, Inc. is the leading provider of training and certification programs, services and products to users of lifting equipment in North America. Through our training and certification programs thousands of operators, inspectors, riggers and trainers have been trained, qualified and certified, all having a positive impact toward reducing accidents. Additionally, supervisors, managers and equipment owners have been provided with the training necessary to establish and improve safety programs on their job sites. Our training and certification programs are offered in two formats, Scheduled and On-Site:

Scheduled Programs

- Qualified Rigger/Signalperson (2 days)
- Advanced Rigger (1 day)
- Rigging Equipment Inspector (2 days)
- Managing Crane Safety (2 days)
- Mobile Crane Operator (3 days)
- Mobile Crane Operator, Hands-On (4 days)
- Mobile Crane Inspector (4 days)
- Small Crane Operator, Hands-On (4 days)
- Tower Crane Operator & Inspector (4 days)
- Overhead Crane Operator & Inspector (4 days)
- Aerial lift Operator & Inspector (3 days)
- Forklift Operator & Inspector (3 days)
- Train the Trainer:
  - Aerial Lift Operator (3 days)
  - Forklift Operator (3 days)
  - Mobile Crane Operator (7 days)
  - Small Crane Operator (5 days)
  - Overhead Crane Operator (4 days)
  - Rigger/Signalperson (4 days)

These programs can be conducted on-site, at your facility and tailored to meet your specific needs and equipment.

Go to www.craneinstitute.com for more information, dates, and prices.

CRANE INSTITUTE OF AMERICA

Crane Simulator Training

All of our Mobile Crane Operator programs that are conducted at our headquarters near Orlando, Florida include training sessions on our full motion crane simulator, which has a real operator's cab, actual controls, and a load moment indicator. Our simulator is so realistic you will think you are actually operating a real crane – and all this without the fear of having an accident.

Benefits of Simulator Training

- Provides an environment where students can learn the art of crane operation without the possibility of an accident.
- Students can get immediate feedback regarding their performance.
- Operating deficiencies can be targeted and corrected by repeating an operation.
- Crane operators can become more effective and efficient in the operation of an actual crane.
- Reduces the cost of taking an operational crane off-line for training purposes.

CRANE INSTITUTE OF AMERICA

Hands-on Training

Our open-enrollment "hands-on" training is available at Crane Institute of America's 5-acre facility near Orlando, Florida. With our proven three-step "building block" approach to operator training, students move from the classroom - to the crane simulator - to the actual operation of the equipment, acquiring the knowledge and practical skills they'll need back on the job. Call 1-800-832-2726 to enroll in any of our programs.

On-Site Training Programs

These training programs are specifically developed for the customer's cranes and equipment and are conducted on-site at the customer's facility. Technical information is presented in a classroom setting followed by fieldwork where students are taught the basic and advanced techniques of actual crane operation, rigging, and inspection. Training can be conducted on all types of cranes and equipment for all levels of personnel.

Our safety, management, inspection and train-the-trainer programs are offered on the following types of cranes and equipment: mobile cranes, overhead and gantry cranes, tower cranes, portal cranes, aerial lifts, forklifts, basic and advanced rigging, scaffolding and fall protection. Certification is available at both scheduled and on-site programs for operators, inspectors, riggers and trainers who meet the requirements.

Contact us at 1-800-832-2726 for a cost proposal.

CRANE INSTITUTE OF AMERICA

Support Services

Legal and Investigative Support

Crane Institute of America employs highly trained experts in Accident Investigation and Reconstruction. As noted authorities in the safety and technical aspects of cranes and rigging, CIA personnel frequently provide expert testimony and litigation support, as well as interpretation of standards and regulations.

Site Safety Assessment

During our on-site visit we will assess your personnel, material handling practices, records and existing documentation for compliance with federal, state and industry requirements. Concluding our evaluation, an exit interview will be conducted followed by a written assessment and recommendations report. A thorough safety analysis of your workplace helps prevent accidents before they happen.

In-House Training Programs

These custom training programs are developed by Crane Institute of America for companies who train their crane operators, riggers and signalpersons in-house. Experienced graphic designers use state-of-the-art software to produce full-color visual training aids which include slides, videos, CD-ROMs and presentation programs. Custom programs also include a student manual, instructor guide, exercises and job aids. Crane Institute of America will assist in the implementation of the program and train and certify your in-house trainers.

Crane Institute of America, Inc. is accredited by the U.S. Department of Labor under Part 1919.

CRANE INSTITUTE OF AMERICA

Accredited Certifications

Crane Institute of America Certification (CIC) offers NCCA nationally accredited certifications which are recognized by OSHA.

Rigger Certifications
- Basic Rigger/Signalperson
- Advanced Rigger

Crane Operator Certifications
- Small Telescoping Boom, under 21 Tons
- Medium Telescoping Boom, 21-75 Tons
- Large Telescoping Boom, 75-300 Tons
- XLarge Telescoping Boom, over 300 Tons
- Lattice Boom Carrier/Crawler, under 300 Tons
- Lattice Boom Carrier/Crawler, over 300 Tons
- Friction Crane
- Articulating Boom
- Digger Derrick
- Service/Mechanic Truck

A Higher Level of Certification

Advantages of CIC Certifications:
- Meets OSHA's type and capacity requirement
- Only one (1) practical exam for up to five (5) levels of certification
- Realistic written and practical exams
- No hidden fees
- Excellent customer service

Recognized

Crane Inspector/Certifier Certifications
- Mobile Cranes
- Tower Cranes
- Overhead Cranes

Accredited

CRANE INSTITUTE OF AMERICA

Handbooks

Mobile Cranes
Companion volume to "Rigging"

The most up-to-date handbook available on mobile cranes. Topics include load charts, pre-operational inspection, working around power lines, operating procedures, hoisting personnel and more. A training program in itself, and excellent preparation for licensing and certification exams.

Forklifts

Pocket-sized and user friendly, this comprehensive handbook is designed to provide operators, managers and safety professionals with the tools to identify and control lift truck related hazards.

Scaffolding Safety

Written to explain and illustrate OSHA's standard on scaffolding, this handbook is a must for all personnel, especially OSHA's required "competent person." Like all our books, the combination of graphics with text makes the information easy to understand and apply.

Fall Protection

This handbook is designed to discuss types, requirements and the application of standard fall protection systems. Using OSHA as a guide, it serves as a model for training personnel in effectively reducing the risks associated with fall hazards in the workplace.

CRANE INSTITUTE OF AMERICA

Field Guides

Crane Institute of America has developed and published a series of high quality field guides on the following topics. Managers and trainers will find these guides ideal for toolbox talks and safety references.

- Crane Setup
- Basic Operating Practices
- Working Cranes Near Power Lines
- Hoisting Personnel by Crane
- Pre-Operational Inspection
- Reductions in Rated Capacity
- How to Use Load Charts

Inspection Aids

- Inspection Certificate Decals
- Sheave Gauges (Standard & Metric)
- Calipers (Stainless Steel)
- Checklists for Annual or Periodic Inspections*

Telescoping Boom Cranes	Vehicle-Mounted Aerial Lifts
Lattice Boom Cranes	Boom-Supported Aerial Lifts
Boom Trucks	Industrial Lift Trucks
Bridge & Gantry Cranes	Rough Terrain Lift Trucks
Jib Cranes & Hoists	Chain Slings
Monorail Crane Systems	Manual Chain Hoists
Tower Cranes	Lever Operated Hoists
Articulating Boom Cranes	Service/Mechanics Trucks

    New Offer: Put your logo on our annual checklist

- Checklists for Pre-Operational Inspections

Telescoping Boom Cranes	Aerial Lifts
Lattice Boom Cranes	Industrial Lift Trucks
Overhead Cranes & Hoists	

CRANE INSTITUTE OF AMERICA

Safety Training DVD's & Videos

- Advanced Tips for Rigging & Lifting
- Rigging & Lifting with Small Hydraulic Cranes*
- Hand Signal Communication
- Working Cranes Near Power Lines
- Hand Signal Communications
- Setup for Safety
- Overhead Crane Safety
- Rigging & Lifting with Mobile Construction Equipment*
- Pre-Op Inspection for Telescoping Boom Cranes
- Pre-Op Inspection for Lattice Boom Cranes
- How to Properly Interpret a Load Chart
- Sliding Boom (Telescoping) Forklift Operating Tech

*Available in Spanish.

Safety Aids

- Safety Reference Cards
  - Crane Safety
  - Hoisting Personnel
  - Rigging
- Hand Signal Cards
  - Overhead Crane
  - Mobile Crane*
  - Tower Crane
- Hand Signal Charts (self-adhesive)
  - Overhead Crane
  - Mobile Crane*

*Now available in Spanish!

CRANE INSTITUTE OF AMERICA

Scale Model Cranes and Lift Equipment
Beautifully crafted, finely detailed, certain to please any collector

Crane Models

- Grove RT 540E
- Grove RT 700E
- Grove TM 1500
- Manitowoc 4100W
- Potain MDT-Tower
- Terex Demag AC 35

Visit us online for a complete list of available models

Terex Demag AC 35
1:50 scale (7 in. long)

Apparrel and more...

Hats
Khaki with Navy trim
Navy with Stone trim

T-shirts
Gray with full color back
White with logo

Polo shirts
Available in various colors

CRANE INSTITUTE OF AMERICA

ASME B30 Standards

We are authorized by ASME as a book dealer to distribute all ASME B30 standards and codes. Individual volumes cover the following:

- B30.1 Jacks
- B30.2 Overhead Cranes (Single or Multiple Girder)
- B30.3 Construction Tower Cranes
- B30.4 Portal, Tower and Pedestal Cranes
- B30.5 Mobile and Locomotive Cranes
- B30.6 Derricks
- B30.7 Base-Mounted Drum Hoists
- B30.8 Floating Cranes and Floating Derricks
- B30.9 Slings
- B30.10 Hooks
- B30.11 Monorail and Underhung Cranes
- B30.12 Handling Loads Suspended from Rotocraft
- B30.13 Storage/Retrieval Machines and Associated Equipment
- B30.14 Side Boom Tractors
- B30.16 Overhead Hoists (Underhung)
- B30.17 Overhead Cranes
- B30.18 Stacker Cranes
- B30.19 Cableways
- B30.20 Below-the-Hook Lifting Devices
- B30.21 Manually Lever-Operated Hoists
- B30.22 Articulating Boom Cranes
- B30.23 Personnel Lifting Systems
- B30.24 Container Cranes
- B30.25 Scrap and Material Handlers
- B30.26 Rigging Hardware
- B30.27 Material Placement Systems
- B30.28 Balance Lifting Units
- B30.29 Self Erecting Tower Crane

Crane Institute of America, Inc.

To order products call **(800) 832-2726** or **(407) 322-6800**,
visit us at **www.craneinstitute.com**,
or fax an order form to **(407) 330-0660**

NOTES

NOTES

NOTES